T0199339

# Animal Body Size

# Animal Body Size

*Linking Pattern and Process across Space, Time, and Taxonomic Group*

EDITED BY FELISA A. SMITH AND S. KATHLEEN LYONS

THE UNIVERSITY OF CHICAGO PRESS CHICAGO AND LONDON

FELISA A. SMITH is professor of biology at the University of New Mexico. S. KATHLEEN LYONS is a research scientist in the Department of Paleobiology at the National Museum of Natural History.

The University of Chicago Press, Chicago 60637
The University of Chicago Press, Ltd., London

Printed in the United States of America
22 21 20 19 18 17 16 15 14 13     1 2 3 4 5

ISBN-13: 978-0-226-01214-8 (cloth)
ISBN-13: 978-0-226-01228-5 (e-book)
DOI: 10.7208/chicago/9780226012285.001.0001

Library of Congress Cataloging-in-Publication Data

Animal body size : linking pattern and process across space, time, and taxonomic group / edited by Felisa A. Smith and S. Kathleen Lyons.
pages ; cm
Includes bibliographical references and index.
ISBN 978-0-226-01214-8 (cloth : alk. paper) — ISBN 978–0–226–01228-5 (e-book) 1. Body size. 2. Variation (Biology) 3. Macroecology. I. Smith, Felisa A. II. Lyons, S. Kathleen.
QL799.A55 2013
591.4′1—dc23

2012042I2

♾ This paper meets the requirements of ANSI/NISO Z39.48–1992 (Permanence of Paper).

# Contents

# Preface

The idea of a book synthesizing patterns of body size across different taxa was an outgrowth of the working group "Body Size in Ecology and Paleoecology: Linking Pattern and Process across Taxonomic, Spatial, and Temporal Scales," supported by the National Center for Ecological Analysis and Synthesis (NCEAS) during 1999–2003. This group, composed of a diverse set of scientists working on different taxa at different taxonomic and scientific scales, had been inspired by and organized during a 1998 Penrose conference aimed at integrating ecology and paleontology. Despite our divergent backgrounds, our common research interests all revolved around body size: Just how similar are large-scale body size patterns across plant and animal species and across evolutionary time? How does the relative magnitude of these important factors change for different taxa? And what mechanisms underlie the body size patterns observed? To address these questions, we compiled what is now a widely cited global database of mammalian body size, distribution, and taxonomy (MOM v.3.1; Smith et al. 2003) as well as several other body size datasets at regional and global scales on plants, birds, and other taxa (*http://www.nceas.ucsb.edu/projects/2182*). Our efforts led to several dozen papers, several organized symposia at meetings of both the Ecological Society of America and the American Society of Mammalogists and, ultimately, an NSF-sponsored Research Coordination Network (RCN) concentrating on macroecological patterns of mammalian body size (IMPPS; http://biology.unm.edu/impps_rcn/). More important, the interactions profoundly changed the scientific approaches, perspectives, and research interests of some of the individual members.

Our original NCEAS group faced a number of daunting challenges. First, ten years ago there was a paucity of comprehensive data on body size that made examining emergent patterns very difficult. Indeed, much of our time and resources went into collecting and analyzing such data. Second, we found that we spoke different "scientific languages"; hence, considerable effort went into figuring out how to integrate divergent taxonomic, hierarchical, and biological perspectives. Because we were scattered around the world and many of us attended different national meetings, there was little opportunity for dialogue outside our working group. Yet we found that the development of interpersonal relationships was very important to overcoming scientific isolationism and creating a productive working relationship. We were fortunate that the funding structure in place at that time for NCEAS working groups led to a total of eight meetings; twice what is supported now. These "extra" meetings allowed us to overcome many of these social and scientific barriers. Third, there was limited institutional and financial support to continue projects that spanned diverse disciplinary and conceptual boundaries beyond the tenure of the working group. Our ability to do so was especially hampered by the rigid structure of universities and funding agencies (such as the NSF). The physical and philosophical segregation of paleoecological from ecological and evolutionary disciplines made it difficult to conduct or fund synthetic work spanning these traditional boundaries. Thus, we encountered substantial obstacles in our efforts to sustain and broaden collaborations.

Over the past ten years, a number of other groups have been formed to examine body size patterns and evolution. These include a recent working group at the National Evolutionary Synthesis Center (NESCent), "Phanerzoioc Body Size Trends in Time and Space," organized by Jonathan Payne, Jennifer Stempien, and Michał Kowalewski, and the NSF-sponsored Research Coordination Network, "Integrating Macroecological Patterns and Process across Scales" (IMPPS; http://biology.unm.edu/impps_rcn/) organized by Felisa Smith, Kate Lyons, and Morgan Ernest. Although the taxonomic scope of these two groups is different, both aim to synthesize emergent organismal and ecological data and patterns across multiple spatial and temporal scales, and both include paleontologists and ecologists, theoreticians and empiricists. These efforts are leading to important new data compilations and contributions. For example, the NESCent group recently published a paper synthesizing body size trends across the entire history of life on earth, a span of some 3.6 billion years (Payne et al. 2009) and our RCN group has examined patterns of mammalian body size for each order on each continent over its evolutionary history (Smith et al. 2010). The continued interest in the patterns of body size evolution over space and time highlights its acknowledged importance in physiology, ecology, and community and ecosystem structure. It also indicates a continued lack of synthetic knowledge about the universality of patterns and processes across taxonomic, temporal, and spatial scales. We sincerely hope our volume will help bridge some of these gaps.

Finally, as is often the case, this volume took much longer to come to fruition than we originally (and perhaps naively!) intended. We thank all the authors for their patience, but especially those authors who met our deadlines, particularly considering that we did not meet them ourselves. Sadly, a series of serious health and other issues impeded our efforts. We hope they agree that the result was worth the delay.

Felisa Smith and Kate Lyons
December 2009
Santa Fe, New Mexico, and Washington, DC

## References

Payne, J. L., A. G. Boyer, J. H. Brown, S. Finnegan, M. Kowalewski, R. A. Krause, S. K. Lyons, C. R. McClain, D. W. McShea, P. M. Novack-Gottshall, F. A. Smith, J. A. Stempien, and S. C. Wang. 2009. "Two-phase increase in the maximum size of life over 3.5 billion years reflects biological innovation and environmental opportunity." *Proceedings of the National Academy of Sciences of the United States of America* 106 (1): 24–27.

Smith, F. A., A. G. Boyer, J. H. Brown, D. P. Costa, T. Dayan, S. K. M. Ernest, A. R. Evans, M. Fortelius, J. L. Gittleman, M. J. Hamilton, L. E. Harding, K. Lintulaakso, S. K. Lyons, C. McCain, J. G. Okie, J. J. Saarinen, R. M. Sibly, P. R. Stephens, J. Theodor, and M. D. Uhen. 2010. "The evolution of maximum body size of terrestrial mammals." *Science* 330 (6008): 1216–1219. doi: 10.1126/science.1194830.

Smith, F. A., S. K. Lyons, S. K. M. Ernest, K. E. Jones, D. M. Kaufman, T. Dayan, P. A. Marquet, J. H. Brown, and J. P. Haskell. 2003. "Body mass of late Quaternary mammals." *Ecology* 84 (12): 3403–3403.

# Acknowledgments

We are extremely grateful to the members of the original National Center for Ecological Analysis and Synthesis (NCEAS) working group "Body Size in Ecology and Paleoecology: Linking Pattern and Process across Taxonomic, Spatial and Temporal Scales" for their interest and input on all things related to body size. Members included John Alroy, Jim Brown, Ric Charnov, Tamar Dayan, Brian Enquist, Morgan Ernest, Liz Hadly, John Haskell, Dave Jablonski, Kate Jones, Dawn Kaufman, Kate Lyons, Brian Maurer, Karl Niklas, Warren Porter, Kaustuv Roy, Felisa Smith, Bruce Tiffney, and Mike Willig. Several of the chapters included sprang directly from our discussions. Funding for this group came from NCEAS, the National Science Foundation (DEB-0072909), the State of California, and the University of California, Santa Barbara; we are grateful for their support. This project was supported in part by the Integrating Macroecological Pattern and Process across Scales (IMPPS) NSF Research Coordination Network (DEB-0541625); this is IMPPS RCN publication #5. Finally, we thank our children (Emma, Rosy, and Kieran) and spouses (Scott and Pete) for their encouragement, patience, and support over the years.

INTRODUCTION

# On Being the Right Size: The Importance of Size in Life History, Ecology, and Evolution

Felisa A. Smith and S. Kathleen Lyons

"For every type of animal there is an optimum size." —J. B. S. Haldane, *"On Being the Right Size"*

Living things vary enormously in body size. Across the spectrum of life, the size of animals spans more than twenty-one orders of magnitude, from the smallest (mycoplasm) at $\sim10^{-13}$ g to the largest (blue whale) at $10^8$ g (fig. I.1, table I.1). We now know that much of this range was achieved in two "jumps" corresponding to the evolution of eukaryotes and metazoans, at 2.1 Ga and 640 Ma, respectively (Payne et al. 2009). Yet the drivers behind these jumps, the factors underlying similarities and differences in body size distributions, and the factors selecting for the "characteristic" or "optimum" size of organisms remain unresolved (Smith et al. 2004; Storch and Gaston 2004).

The study of body size has a long history in scientific discourse. Some of our earliest scientific treatises speculate on the factors underlying the body mass of organisms (e.g., Aristotle 347–334 B.C.). Many other eminent scientists and philosophers, including Galileo Galilei, Charles Darwin, J. B. S. Haldane, George Gaylord Simpson, and D'Arcy Thompson, have also considered why organisms are the size they are and the consequences of larger or smaller size. As Galileo stated, "Nature cannot produce a horse as large as twenty ordinary horses or a giant ten times taller than an ordinary man unless by miracle or by greatly altering the proportions of his limbs and especially of his bones" (Galileo 1638). The

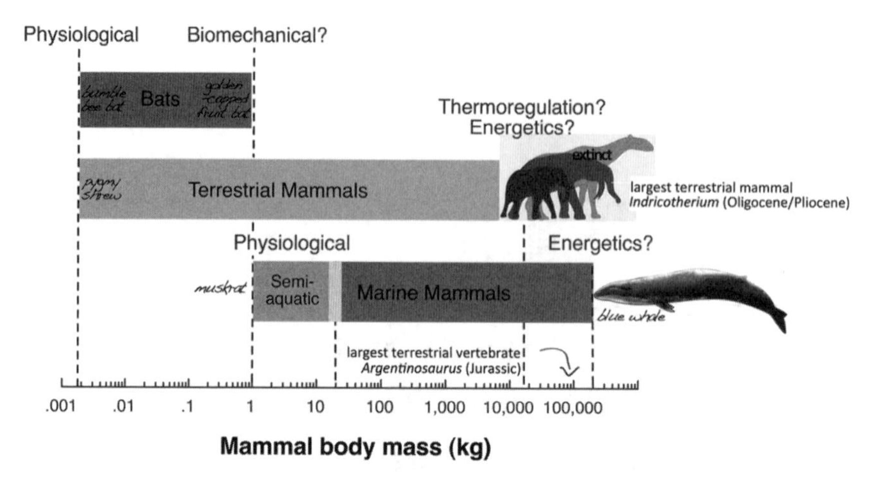


FIGURE I.1. Factors influencing the minimum and maximum size of mammals. Note that the minimum size of ~1.8 g is represented by both volant and nonvolant mammals; aquatic mammals have a much larger minimum body size, which appears to be set by the thermoregulatory demands of living in an aquatic environment (see text). The largest terrestrial mammal, *Indricotherium*, reached masses reportedly in excess of 12–15 tons. Interestingly, this is about an order of magnitude smaller than the largest terrestrial dinosaurs and could reflect a difference between endothermic and exothermic animals if resources limit size in terrestrial environments.

fascination with body size stems not only from the ability to clearly characterize it, but also from the fact that it so clearly matters.

Over the past few decades, considerable research has gone into understanding the physiological consequences of being a certain size. This was inspired in part by the almost concurrent publication of three seminal books on body size in the mid-1980s: Peters (1983), Calder (1984), and Schmidt-Nielsen (1984). Thanks to these and other works (e.g., Kleiber 1932), we now know just how many fundamental physiological, ecological, and evolutionary factors scale allometrically (i.e., $y = x^b$, where $y =$ represents some organismal trait, $x$ represents body mass and $b$, an exponent that is generally less than 1) with mass. These include fecundity, energetic requirements, diet, territory and home range size, longevity, and even extinction rates (Bourliere 1975; Niklas 1994).

There has also been a recent and exciting development of mechanistic mathematical models, rooted in specific aspects of individual anatomy and physiology. These models attempt to bridge the gap between body size patterns that are present across differing temporal and spatial

TABLE I.1 **The Range of Body Size of Various Taxa**

| Taxon | Smallest | | Largest | | Range of Size (Orders of Magnitude) |
|---|---|---|---|---|---|
| | Organism | Size | Organism | Size | |
| Class Mammalia (aquatic) | *Enhydra lutris* (sea otter) | ~27 kg | *Balaenoptera musculus* (blue whale) | ~180 tons | 8 (mass) |
| Class Mammalia (terrestrial) | *Suncus etruscus* (pygmy shrew) | ~1.8 g | *Indricotherium transouralicum* (extinct) | 12–15 tons | 7 (mass) |
| Class Reptilia (marine) | *Keichousaurus* | 0.454 kg | *Hainosaurus* | ~15 m, ~15 tons | 4 (mass) |
| Class Reptilia (terrestrial; nondinosaurs) | *Spaerodactylus ariasae* (gecko) | 16 mm | *Sarcosuchus imperator* (extinct) | ~12 m, ~13.6 tons | 3 (length) |
| Class Reptilia (turtles) | *Homopus signatus* (speckled padloper tortoise) | 8 cm | *Archelon ischyros* (extinct) | 4.84 m, 2200 kg | 2 (length) |
| Class Reptilia (dinosaurs) | *Archiornis* | 34 cm, 110 g | *Argentiosaurus* (extinct) | ~80–90 tons | 6 (mass) |
| Class Aves | *Mellisuga helenae* (bee hummingbird) | 6.2 cm, ~1.8 g | *Struthio camelus* (North African ostrich) | ~2.75m, ~156 kg | 2 (length) |
| Class Amphibia (frogs) | *Eleutherodactylus Iberia* (Monte Iberia eleuth) | ~9.8 mm | *Conraua goliath* (goliath frog) | ~32 cm, 3.3 kg | ~2 (length) |
| Superclass Osteichthyes | *Paedocypris progenetica* | ~7.9 mm long | *Rhincodon typus* (whale shark) | 12.6 m long | 4 (length) |
| Class Amphibia (all) | *Brachycephalus didactylus* (Brazilian golden frog) | 9.8 mm | *Prionosuchus* (extinct) | 9 m | 3 (length) |
| Class Bivalvia | *Sphaeriidae* | 0.2 mm | *Platyceramus platinus* (extinct) | 3 m | 4 (length) |
| Class Gastropoda | *Ammonicera rota* | 0.05 cm | *Syrinx aruanus* | 91 cm, 18 kg | 3 (length) |
| Class Cephalopoda | *Idiosepius notoides* (pygmy squid) | 7 mm | *Mesonychoteuthis hamiltoni* (colossal squid) | 13 m, 494 kg | 4 (length) |
| Class Trilobita | *Ctenophyge ceciliae* (extinct) | 3 mm | *Isotelus rex (extinct)* | 720 mm | 3 (length) |
| Division Angiospermae | *Salix herbacea* (dwarf willow) | 1–6 cm tall | *Sequoiadendron giganteum* (giant sequoia) | 83.8 m tall | 3 (length) |
| Class Arachnida | *Patu marplesi* (Samoan moss spider) | 0.3 mm | *Theraphosa blondi* (goliath bird-eating spider) | 28 cm, 170 g | 4 (length) |
| Class Insecta | *Nanosella fungi* (feather-winged beetle) | 0.25 mm | *Goliathus goliatus* (goliath beetle) | >110 mm | 3 (length) |
| Domain Bacteria | *Mycoplasma genitalium* | ~200 nm | *Epulopiscium fishelsoni* | 0.7 mm | 5 (length) |

*Note*: Mass scales as the cube of length for roughly cylindrical organisms. Thus, a difference of one order of magnitude in length is approximately equal to a three-order difference in mass.

scales and have shown that animals and plants share many similar allometric scaling relationships (e.g., Brown et al. 1993; West et al. 1997; Enquist et al. 1998; Maurer 1998; West et al. 1999). This is an important insight, as it suggests that across diverse groups of organisms (i.e., plants and animals) body size distributions may in fact not only be similar but also predictable. Yet how the complex and dynamic interactions among intrinsic structure and function, environment, and historical and/or phylogenetic evolution result in particular body sizes remains unclear.

Paleontologists, ecologists, and comparative evolutionary biologists have also extensively studied body size. The detailed analysis of size patterns has lead to the formation of several well-supported large-scale biogeographic and temporal "rules," such as Bergmann's, Cope's, and Foster's rules, and the plant self-thinning law (Bergmann 1847; Cope 1887; Yoda et al. 1963; Foster 1964). Just how pervasive these spatial and temporal phenomena are, however, is still the subject of considerable debate (Weller 1987; Lonsdale 1990; Jablonski 1997; Alroy 1998). With few exceptions (e.g., Brown and Maurer 1986; Jablonski 1993; Brown 1995; Jablonski and Raup 1995; Smith et al. 1995; Smith and Betancourt 1998) biogeographic and temporal patterns have been studied separately, with ecologists focusing on the former, and paleoecologists on the latter (e.g., Cope 1887; Mayr 1956; Stanley 1973; MacFadden 1987; Damuth and MacFadden 1990, and references therein; Morgan et al. 1995; Hadly 1997; Jablonski 1997; Alroy 1998; Enquist et al. 1998). Likewise, although comparative biologists have made admirable progress examining trait evolution in contemporary taxa (e.g., Harvey and Purvis 1991; Purvis and Harvey 1997; Bininda-Emonds et al. 2001), they have had difficulty linking ecological processes with evolutionary ones and often ignore the fossil record. There is little integration across the divergent scales studied by ecologists, comparative evolutionary biologists, and paleoecologists, and limited attempts have been made to span taxonomic or other boundaries.

To some extent, these issues have stemmed at least partially from difficulties in assembling appropriate data. Several earlier studies investigated the influence of environmental conditions and evolutionary constraints on size, for example, but were limited in geographic or taxonomic scope (Hutchinson and MacArthur 1959; May 1978, 1986; Brown and Nicoletto 1991; Pagel 1999; but see Alroy 1998; Harvey 2000; Blackburn and Gaston 2001). Other authors compared body size and/or life history traits across continents but focused on particular orders where data were available, utilized a subset of taxa, or conducted analyses at the generic or familial level (e.g., Read and Harvey 1989; Maurer et al. 1992; Kappeler and Heymann 1996). In particular, the impact of phylogeny on the pattern and similarity of body size remains underexplored, especially across different taxa and scales. Methods for exploring the phylogenetic signal in traits, however, are becoming more robust (e.g., Freckleton et al. 2002; Blomberg et al. 2003) and a number of synthetic datasets have become available (e.g., Smith et al. 2003, among others; Jones et al. 2009).

Thus, there are a number of profoundly important questions that remain unaddressed by any subdiscipline of biology. First, how do the complex interactions between organic structure and function, environment, and historical and/or phylogenetic evolution engender particular body sizes, and, second, how do these interactions evoke the apparently remarkably consistent body size patterns seen across taxa, space, and time? A potentially powerful hypothesis is that size frequency distributions are similarly skewed to the right or to the left because the organisms contributing to these distributions share similar trophic or life history traits despite their other phyletic differences. Do organisms of similar size demonstrate similarities in life history traits? If so, what are the relative contributions of phylogenetic autocorrelation, environmental factors, and architectural limitations? Do spatially averaged distributions have a different form from those of temporally averaged distributions? Do emergent statistical patterns also exist across time, and if so, how consistent have they been over time? How similar are body size patterns across plant and animals species? How do these important factors interact for different taxa? Do certain sizes make clades more likely to speciate by either decreasing the chances of extinction or increasing the likelihood of speciation? Several studies have examined the influence of body size in influencing species richness (e.g., Dial and Marzluff 1988; Gittleman and Purvis 1998; Orme et al. 2002), but more comprehensive tests across different lineages are required. Last, what are the mode and tempo of body size change through evolutionary and ecological time, and across different taxa?

All of the papers here take a macroecological approach to examine the patterns and underlying causal mechanisms of body size. That is, they emphasize description and explanation of the emergent statistical properties of large numbers of ecological "particles," be they individuals, populations, or species (Brown 1995). This is not surprising:

body size has been a key variable for many macroecological studies. Indeed, the use of a macroecological approach is increasingly common when addressing fundamental questions in ecology and paleoecology at large spatial and temporal scales (Smith et al. 2008). The book is divided into two parts: "Body Size Patterns across Space and Time" and "Mechanisms and Consequences Underlying Body Size Distributional Patterns." Within each section, we have included papers reflecting different taxonomic, hierarchical, and/or scientific perspectives.

While our volume does not provide answers to all the intriguing questions raised by the past few decades of research on body size, it does address many of these issues. For example, several papers (e.g., those by Gaston and Chown; Nekola et al.; Maurer; Lyons and Smith; Ernest; and Maurer and Marquet) deal with the fundamental interactions between body size and population, community, and/or ecosystem structure and function. Others explore the role of life history (Safi et al.; Brown et al.) or modes of life (Smith et al.) in patterns of body size. Divergent scientific perspectives are presented, including explicit phylogenetic or taxonomic (Gaston and Chown; Maurer; Safi et al.; Smith et al.), theoretical (Brown et al.) and paleoecological approaches (Lyons and Smith). By compiling these contributions in one volume, we have attempted to highlight some of the patterns common across spatial, temporal, and/or taxonomic scales. However, despite the recent resurgence of interest in understanding and explaining emergent patterns of body size at varying scales, there is still no consensus on the drivers or triggers underlying these processes. We hope this volume will help inspire new research on some of these big, important, and still unanswered questions.

## References

Alroy, J. 1998. "Cope's rule and the dynamics of body mass evolution in North American fossil mammals." *Science* 280 (5364): 731–734.

Aristotle. 347–334 B.C. *De partibus animalium*. In *The complete works of Aristotle*, edited by J. Barnes, translated by L. Dittmeyer (1907; rev. ed., Princeton: Pinceton University Press, 1984).

Bergmann, C. 1847. "Ueber die Verhältnisse der Wärmeökonomie der Thiere zu ihrer Grösse." *Göttinger Studien* 3:595–708.

Bininda-Emonds, O. R. P., J. L. Gittleman, and C. K. Kelly. 2001. "Flippers versus feet: Comparative trends in aquatic and non-aquatic carnivores." *Journal of Animal Ecology* 70:386–400.

Blackburn, T. M., and K. J. Gaston. 2001. "Local avian assemblages as random draws from regional pools." *Ecography* 24 (1): 50–58.

Blomberg, S. P., T. Garland, and A. R. Ives. 2003. "Testing for phylogenetic signal in comparative data: Behavioral traits are more labile." *Evolution* 57: 717–745.

Bourliere, F. 1975. "Mammals, small and large: The ecological implications of size." In *Small mammals: Their productivity and population dynamics*, edited by F. B. Golley, K. Petrusewicz, and L. Ryszkowski. Cambridge: Cambridge University Press.

Brown, J. H. 1995. *Macroecology*. Chicago: University of Chicago Press.

Brown, J. H., P. A. Marquet, and M. L. Taper. 1993. "Evolution of body size: Consequences of an energetic definition of fitness." *American Naturalist* 142 (4): 573–584.

Brown, J. H., and B. A. Maurer. 1986. "Body size, ecological dominance and Cope's rule." *Nature* 324 (6094): 248–250.

Brown, J. H., and P. F. Nicoletto. 1991. "Spatial scaling of species composition: Body masses of North American land mammals." *American Naturalist* 138 (6): 1478–1512.

Calder, W. A. 1984. *Size, function, and life history*. Cambridge, MA: Harvard University Press.

Cope, E. D. 1887. *The origin of the fittest*. New York: Appleton.

Damuth, J., and B. J. MacFadden. 1990. *Body size in mammalian paleobiology: Estimation and biological implications*. New York: Cambridge University Press.

Dial, K. P., and J. M. Marzluff. 1988. "Are the smallest organisms the most diverse?" *Ecology* 69:1620–1624.

Enquist, B. J., J. H. Brown, and G. B. West. 1998. "Allometric scaling of plant energetics and population density." *Nature* 395 (6698): 163–165.

Foster, J. B. 1964. "Evolution of mammals on islands." *Nature* 202:234–235.

Freckleton, R. P., P. H. Harvey, and M. Pagel. 2002. "Phylogenetic analysis and comparative data: A test and review of evidence." *American Naturalist* 160 (6): 712–726.

Galileo, G. 1638. "Discorsi e dimostrazioni matematiche, intorno à due nuove scienze." In *Mathematical discourses and demonstrations, relating to two new sciences*, translated by Henry Crew and Alfonso de Salvio (Leiden: Elsevier, 1914).

Gittleman, J. L., and A. Purvis. 1998. "Body size and species-richness in carnivores and primates." *Proceedings of the Royal Society B: Biological Sciences* 265:113–119.

Hadly, E. A. 1997. "Evolutionary and ecological response of pocket gophers (*Thomomys talpoides*) to late-Holocene climatic change." *Biological Journal of the Linnean Society* 60:277–296.

Harvey, P. H. 2000. "Why and how phylogenetic relationships should be incorporated into studies of scaling." In *Scaling in biology*, edited by J. H. Brown and G. B. West. Oxford: Oxford University Press.

Harvey, P. H., and A. Purvis. 1991. "Comparative methods for explaining adaptations." *Nature* 351:619–624.

Hutchinson, G. E., and R. H. MacArthur. 1959. "A theoretical ecological model of size distributions among species of animals." *American Naturalist* 93:117–125.

Jablonski, D. 1993. "The tropics as a source of evolutionary novelty through geological time." *Nature* 364 (6433): 142–144.

———. 1997. "Body-size evolution in Cretaceous molluscs and the status of Cope's rule." *Nature* 385 (6613): 250–252.

Jablonski, D., and D. M. Raup. 1995. "Selectivity of end-Cretaceous marine bivalve extinctions." *Science* 268 (5209): 389–391.

Jones, K. E., J. Bielby, M. Cardillo, S. A. Fritz, J. O'Dell, C. D. L. Orme, K. Safi, W. Sechrest, E. H. Boakes, C. Carbone, C. Connolly, M. J. Cutts, J. K. Foster, R. Grenyer, M. Habib, C. A. Plaster, S. A. Price, E. A. Rigby, J. Rist, A. Teacher, O. R. P. Bininda-Emonds, J. L. Gittleman, G. M. Mace, and A. Purvis. 2009. "PanTHERIA: A species-level database of life-history, ecology and geography of extant and recently extinct mammals." *Ecology* 90:2648.

Kappeler, P. M., and E. W. Heymann. 1996. "Nonconvergence in the evolution of primate life history and socio-ecology." *Biological Journal of the Linnean Society* 59:297–326.

Kleiber, M. 1932. "Body size and metabolism." *Hilgardia* 6:315–351.

Lonsdale, W. M. 1990. "The self-thinning rule—dead or alive?" *Ecology* 71:1373–1388.

MacFadden, B. J. 1987. "Fossil horses from "Eohippus" (*Hyracotherium*) to *Equus*: Scaling, Cope's law and the evolution of body size." *Paleobiology* 12:355–369.

Maurer, B. A. 1998. "The evolution of body size in birds. I. Evidence for nonrandom diversification." *Evolutionary Ecology* 12 (8): 925–934.

Maurer, B. A., J. H. Brown, and R. D. Rusler. 1992. "The micro and macro in body size evolution." *Evolution* 46 (4): 939–953.

May, R. M. 1978. "The dynamics and diversity of insect faunas." In *Diversity of insect faunas*, edited by L. A. Mound and N. Waloff. New York: Blackwell Scientific Publications.

———. 1986. "The search for patterns in the balance of nature: Advances and retreats." *Ecology* 67:1115–1126.

Mayr, E. 1956. "Geographical character gradients and climatic adaptation." *Evolution* 10:105–108.

Morgan, M. E., C. Badgley, G. F. Gunnell, P. D. Gingerich, and J. W. Kappelman. 1995. "Comparative paleoecology of paleogene and neogene mammalian faunas: Body-size structure." *Palaeogeography, Palaeoclimatology, Palaeoecology* 115:287–317.

Niklas, K. J. 1994. *Plant allometry: The scaling of form and process.* Chicago: University of Chicago Press.

Orme, C. D. L., N. J. B. Isaac, and A. Purvis. 2002. "Are most species small? Not within species-level phylogenies." *Proceedings of the Royal Society B: Biological Sciences* 269 (1497): 1279–1287. doi: 10.1098/rspb.2002.2003.

Pagel, M. D. 1999. "Inferring the historical pattern of biological evolution." *Nature* 401:877–884.

Payne, J. L., A. G. Boyer, J. H. Brown, S. Finnegan, M. Kowalewski, R. A. Krause, S. K. Lyons, C. R. McClain, D. W. McShea, P. M. Novack-Gottshall, F. A. Smith, J. A. Stempien, and S. C. Wang. 2009. "Two-phase increase in the maximum size of life over 3.5 billion years reflects biological innovation and environmental opportunity." *Proceedings of the National Academy of Sciences of the United States of America* 106 (1): 24–27.

Peters, R. H. 1983. *The ecological implications of body size.* Cambridge: Cambridge University Press.

Purvis, A., and P. H. Harvey. 1997. "The right size for a mammal." *Nature* 386:332–333.

Read, A. F., and P. H. Harvey. 1989. "Life-history differences among the Eutherian radiations." *Journal of Zoology* 219:329–353.

Schmidt-Nielsen, K. 1984. *Scaling, why is animal size so important?* Cambridge: Cambridge University Press.

Smith, F. A., and J. L. Betancourt. 1998. "Response of bushy-tailed woodrats (*Neotoma cinerea*) to late Quaternary climatic change in the Colorado Plateau." *Quaternary Research* 47:1–11.

Smith, F. A., J. L. Betancourt, and J. H. Brown. 1995. "Evolution of body-size in the woodrat over the past 25,000 years of climate-change." *Science* 270 (5244): 2012–2014.

Smith, F. A., J. H. Brown, J. P. Haskell, S. K. Lyons, J. Alroy, E. L. Charnov, T. Dayan, B. J. Enquist, S. K. M. Ernest, E. A. Hadly, K. E. Jones, D. M. Kaufman, P. A. Marquet, B. A. Maurer, K. J. Niklas, W. P. Porter, B. Tiffney, and M. R. Willig. 2004. "Similarity of mammalian body size across the taxonomic hierarchy and across space and time." *American Naturalist* 163 (5): 672–691.

Smith, F. A., S. K. Lyons, S. K. M. Ernest, and J. H. Brown. 2008. "Macroecology: More than the division of food and space among species on continents." *Progress in Physical Geography* 32 (2): 115–138.

Smith, F. A., S. K. Lyons, S. K. M. Ernest, K. E. Jones, D. M. Kaufman, T. Dayan, P. A. Marquet, J. H. Brown, and J. P. Haskell. 2003. "Body mass of late Quaternary mammals." *Ecology* 84 (12): 3403–3403.

Stanley, S. M. 1973. "An explanation for Cope's rule." *Evolution* 27:1–26.

Storch, D., and K. J. Gaston. 2004. "Untangling ecological complexity on different scales of space and time." *Basic and Applied Ecology* 5 (5): 389–400.

Weller, D. E. 1987. "A reevaluation of the −3/2 power rule of plant self-thinning." *Ecological Monographs* 57:23–43.

West, G. B., J. H. Brown, and B. J. Enquist. 1997. "A general model for the origin of allometric scaling laws in biology." *Science* 276 (5309): 122–126.

———. 1999. "A general model for the structure and allometry of plant vascular systems." *Nature* 400 (6745): 664–667.

Yoda, K., T. Kira, H. Ogawa, and K. Hozumi. 1963. "Self-thinning in overcrowded pure stands under cultivated and natural conditions." *Journal of Biology of the University Osaka City* 14:107–129.

PART I

# Body Size Patterns across Space and Time

CHAPTER ONE

# Macroecological Patterns in Insect Body Size

Kevin J. Gaston and Steven L. Chown

## Introduction

Imagine a continuum representing the level of understanding of macroecological patterns in body size. Toward one extreme lie groups like birds and mammals, for which many of these patterns have been well documented, their mechanisms have been extensively explored (albeit not necessarily resolved), and the level of understanding is being driven vigorously forward. Toward the other lie the insects, for which many patterns are poorly documented, the mechanisms remain little explored, and the level of understanding is often rather limited. The insects are found in this position in major part for two reasons. The first is the extraordinarily large numbers of species. Best estimates are that there are 4–5 million extant species of insects, of which, depending in marked degree on the extent of synonymy, perhaps 0.75–1 million have been formally taxonomically described (previous much higher estimates of overall species richness have largely been discounted—Gaston and Hudson 1994; Gauld and Gaston 1995; May 2000; Ødegaard et al. 2000; Dolphin and Quicke 2001; Novotný et al. 2002; Chapman 2009). The second reason is that the sampling of this diversity has been extremely heterogeneous in space, with, for example, the geographic distribution of those species that have been taxonomically described not conforming with predicted patterns of actual richness (Gaston 1994). Faced with such issues, the best that can usually be done is to tackle the macroecological patterns in particular, less speciose higher taxa, such as tribes or families, and often then only for limited regions of the world. Nonetheless, given that insects are likely to constitute 60% or more of all extant eukaryote species (and a substantial proportion of all the eukaryotic species that have ever existed), any claim to a solid general understanding of the macroecological patterns of body size requires more than an acknowledgment that studies of these issues are difficult for insects (Diniz-Filho et al. 2010). This is particularly so given that examples from the insects are widely employed to support general arguments about the determinants of macroecological patterns in body size, such as the limitations on size posed by structural constraints of body plans, and the need for a paleontological perspective on present-day trends.

In this chapter, we provide an overview of the current understanding of the form of macroecological patterns in insect body size, with particular emphasis on global patterns, patterns through time, and patterns through space. Throughout, we concentrate foremost on patterns at large geographic scales, rather than those in the body sizes of particular local assemblages, although the latter have received substantially more attention (e.g., Janzen 1973; Morse et al. 1988; Basset and Kitching 1991; Basset 1997; Siemann et al. 1999b; Hodkinson and Casson 2000; Krüger and McGavin 2000; Ulrich 2004). The proximate physiological and cellular mechanisms that regulate body size are also not dealt with here. They are the subject of a large literature (e.g., Nijhout 2003; Davidowitz et al. 2005; Edgar 2006; Nijhout et al. 2006; Mirth and Riddiford 2007; Whitman and Ananthakrishnan 2009; see also Chown and Gaston 2010), and understanding in the area is developing rapidly (e.g., Nijhout et al. 2010; Stillwell and Davidowitz 2010a, 2010b; Callier and Nijhout 2011). Nonetheless, we touch on these mechanisms where relevant in a macroecological context.

## Global Patterns

### *The Smallest and the Largest*

The variation in body size exhibited by extant insects is marked. At the level of the individual, the increase in size from egg to adult can be as much as 43,000% (Klok and Chown 1999), and may be larger (see also Whitman 2008). Typically, this variation has substantial implications for the resources used by growing individuals, often entailing fundamentally different approaches to acquisition at different life stages (e.g., Gaston et al. 1991), with the most extreme changes being represented by alterations both in size and habitat in aquatic insects with aerial adult stages. Ontogenetic changes in insect body size also alter the kinds of predators that feed on them. In anomalous emperor moth *Imbrasia belina* caterpillars (or mopane worms), the early instars are eaten by small predators such as insects and gleaning birds, the later instars by a variety of insectivorous birds, and the last instar by reptiles, mammalian carnivores, and humans (Gaston et al. 1997).

Variation in size among individuals of a given species at a given life stage—constrained intraspecific variation (sensu Spicer and Gaston 1999)—can also be substantial. At least some of this variation is a consequence of food availability to juvenile stages (Emlen 1997b; Emlen and Nijhout 2000; Peat et al. 2005; Whitman 2008). In turn, size variation may have substantial effects on fecundity (Honěk 1993), resource allocation trade-offs, and mating strategies. The latter are particularly well known in beetles (Emlen and Nijhout 2000) but are also found in exopterygotes (e.g., Tomkins and Brown 2004). In many social species, size variation is determined by caste membership and variation within a given caste (Oster and Wilson 1978; Fraser et al. 2000; Hughes et al. 2003). More extreme cases include leaf-cutter ants *Atta* spp. and army ants *Eciton hamatum*, where workers range from 0.0025 to 0.0206 g and from 0.0017 to 0.027 g, respectively (Feener et al. 1988; Roces and Lighton 1995). Considerable intraspecific plasticity means that variation in the size of individual species can perhaps more closely approach variation in the average sizes of different species within some taxonomic groups of insects than is the case for taxa of many other kinds of organisms. In other words, the extent to which, say, the mean or median size of individuals of a given life stage adequately characterizes the body size of a species for the purposes of comparative analyses is less secure for insects than it is for many other taxa. Using average sizes instead of the size of the individual of the given species of interest can result in substantially different, and perhaps incorrect, interpretations of the relationships between body size and ecosystem properties (Cohen et al. 2005).

Across the insects as a whole, the smallest species has been argued to be the mymarid egg-parasitic wasp *Dicopomorpha echmepterygis*, males of which are wingless and measure as little as 139 μm, with females being approximately 40% larger (Gahlhoff 1998). Given that this species was not described until 1997, it is not unlikely that there are yet smaller species still to be found. Although there is much debate, five beetle species have been argued to contend the claim to be the largest insect in terms of measurable bulk: the cerambycid *Titanus giganteus* (167 mm), the elephant beetles *Megasoma elephas* (137 mm) and *M. actaeon* (135 mm), and the goliath beetles *Goliathus goliatus* and *G. regius* (110 mm) (Williams 2001). This gives a range of body lengths for adult insects of three orders of magnitude. Across the beetles alone, species body lengths may vary to a similar extent, with the feather-winged beetles (Ptiliidae) being as small as 250 μm (Gahlhoff 1998).

The reliance here on length, rather than mass, as a measure of insect body size reflects a general practical constraint. Collection and storage methods for insects often limit opportunities for the direct determination of fresh body masses, and most studies of macroecological patterns in insect body size have thus employed measures of "characteristic linear dimensions" (e.g., body length, forewing length, wingspan). Where body masses have been used, these are commonly derived from general allometric relationships with these dimensions (for such relationships see, e.g., Rogers et al. 1976; Gowing and Recher 1984; Sample et al. 1993; Benke et al. 1999; Kaspari and Weiser 1999; Mercer et al. 2001; Powell and Franks 2006), rather than direct measurement. These relationships typically have substantial variance about them, unless limited to taxonomically rather narrow groups of species, as might be obvious from the considerable variation in body form of higher taxa that have similar body masses (e.g., Bartholomew and Casey 1978). Nonetheless, direct body mass measurements for insects are not uncommon, and it is clear that there is substantial interspecific mass variation across the group. It is at least six orders of magnitude, ranging from the thrips *Apterothrips apteris* on Marion Island at 0.00004 g (Mercer et al. 2001) to the giant weta *Deinacrida heteracantha*, with females weighing up to 71 g (Whitman 2008). However, globally the range is probably seven orders of magnitude, given the small size of mymarid wasps. This range is greater than the approximately four orders of magnitude spanned by birds (Blackburn and Gaston 1994), similar to that of mammals (Smith et al. 2004) but smaller than that of fish, which range from the likely milligram masses of *Paedocypris progenetica* (Kottelat et al. 2006) to the basking and whale sharks, which weigh several thousands of kilograms, so spanning eight or perhaps more orders of magnitude.

The implications of interspecific size variation for physiological and life history traits have been explored in a wide variety of studies. They include changes in chemical composition (Woods et al. 2003; Lease and Wolf 2011), biochemistry (Darveau et al. 2005b; Darveau et al. 2005a), structural features (Lease and Wolf 2010), thermal physiology (Willmer and Unwin 1981; Stevenson 1985), flight performance (Stone and Willmer 1989; Dudley 2000a; Darveau et al. 2005b), metabolic rate and gas exchange characteristics (Chown et al. 2007; White et al. 2007), locomotion speed, costs of transport, and ability to transit different landscapes (Kaspari and Weiser 1999; Chown and Nicolson 2004), preferred microclimates (Kaspari 1993), food intake rates (Reichle 1968), resource use (Kirk 1991; Novotný and Wilson 1997), host specificity (Wasserman and Mitter 1978; Loder et al. 1997; Novotný and Basset 1999), contest competitive ability (Heinrich and Bartholomew 1979), egg size (García-Barros and Munguira 1997), development time (Honěk 1999), and intrinsic rates of increase (Gaston 1988a). In some cases, such as interspecific size variation in metabolic rate, development rate, and mortality, there is likely to be considerable feedback, so that body size is as much a function of these variables as they are of body size at the intraspecific level (see Kozłowski and Weiner 1997; Kozłowski and Gawelczyk 2002). In combination these effects will determine optimum body size and eventually in turn the interspecific relationships.

### *Conservatism of Body Size*

Along with a number of life history traits, the body size of organisms tends in general to be highly phylogenetically conserved. Among the insects, this is also true. In physiological traits, much of the variance is partitioned at the family and genus levels (Chown et al. 2002). Using body masses from the same database used to examine variance partitioning in physiological traits (and including only those genera for which data were available for two or more species), taxonomic orders account for approximately 3% of the variation, families for 38%, genera for 39%, and the remaining variation is partitioned at the species level (Chown and Gaston 2010). Studies of individual orders in particular regions have also found substantial variation being partitioned among higher-level taxa (Loder 1997; Brändle et al. 2000). This partitioning makes intuitive sense, given that within orders such as the Orthoptera, Coleoptera, Hemiptera, Hymenoptera, and Lepidoptera, species take on a wide range of sizes (fig. 1.1), while within a given family size ranges are smaller. However, it

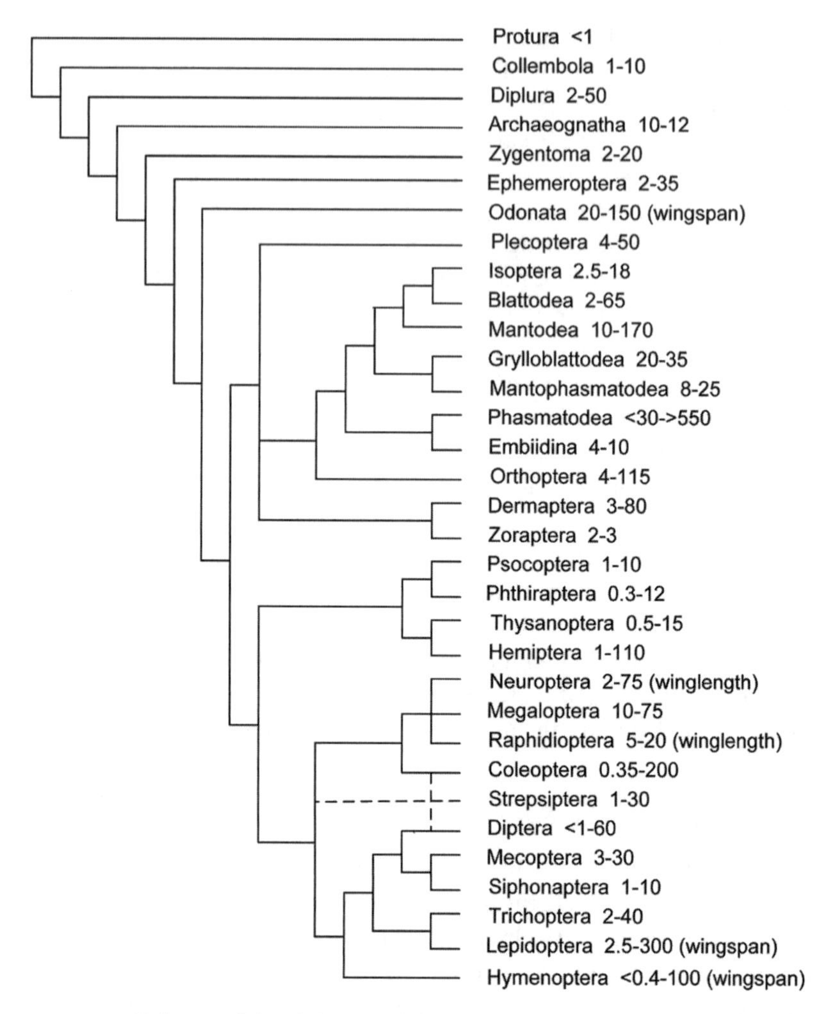


FIGURE 1.1. Estimates of the minimum and maximum adult body lengths (or in some instances wingspan or wing length; all in mm) of species in different insect orders and close relatives, superimposed on a cladogram of their postulated relationships based on combined morphological and nucleotide sequence data. Broken lines indicate uncertain relationships. Redrawn from Chown and Gaston (2010).

also seems likely that as more mass data become available, the variance will be partitioned further toward the generic level, but perhaps more so for the Coleoptera than for any other major order. In the beetles, families tend to be speciose with substantial interspecific size variation (e.g., Carabidae, Curculionidae, Elateridae, Scarabaeidae), while genera tend to be more similar in size, as has long been remarked for carabid beetles (den Boer 1979).

### *Body Size Distributions*

INTRASPECIFIC. The body size distributions of the individuals of particular species have been surprisingly poorly documented for insects, with the exception of the ants and other social Hymenoptera, where frequency distributions of some, usually linear, measure of size are often provided to help understand the causes, consequences, and evolution of the caste distribution function (Oster and Wilson 1978; Hughes et al. 2003; Yang et al. 2004; Schöning et al. 2005; Billick and Carter 2007). At present, although many studies do not highlight the fact that they contain such data, intraspecific body size frequency distributions are readily available for approximately sixty species, representing seven orders and twenty-one families, a tiny portion of insect diversity (Gouws et al. 2011). Nonetheless, several generalities emerge. Among the nonsocial insects, intraspecific body size frequency distributions of both mass and linear measurements are either normally distributed or slightly right skewed. Bimodality seems relatively uncommon. Nonetheless, owing to allometry, substantial variation in the shape of the distributions can exist among linear measures depending on what trait is quantified. In consequence, care must be exercised in selecting the trait to be used for linear measurements and in reporting such distributions (Whitman 2008; Gouws et al. 2011). Among the social insects (mostly ants, but also a termite and wasp species) intraspecific body size frequency distributions appear typically to be right skewed or bimodal (Gouws et al. 2011). However, these authors echo Oster and Wilson's (1978) remarks that perhaps only 15% of ant genera show size polymorphism, and that most species are monomorphic. Thus, most intraspecific body size frequency distributions are probably approximately normal, with perhaps some tendency to show a positive skew. Despite these general patterns, the form of intraspecific body size frequency distributions (rather than their central tendency, which we discuss below) need not be static, and may change through space and time for several reasons (Waddington et al. 1986; Lounibos 1994; Evans 2000; Roulston and Cane 2000; Yang et al. 2004; Peat et al. 2005; Couvillon et al. 2010).

Although the frequency distributions presented by Gouws et al. (2011) and others do not immediately reveal sexual size dimorphism (SSD), males are usually smaller than females in insects (Teder and Tammaru 2005; Whitman 2008; Stillwell et al. 2010). This may be true of whole higher taxa, such as the extant Mantophasmatodea (Klass et al. 2003), where the difference is substantial. In *Karoophasma biedouwensis* males weigh 54.9 ± 8.3 mg, and females 128.6 ± 7.6 mg (Chown et al. 2006). The opposite pattern is not uncommon, however (Stillwell et al. 2010). Moreover, in a range of species there is substantial male size polymorphism (in polymorphic ants and some other social Hymenoptera it is the females that are variable; Emlen and Nijhout 2000). Intraspecific variation in SSD may well be a consequence of adaptive canalization of reproductive traits (Fairbairn 2005). At the interspecific level, SSD typically increases with size when males are the larger sex but declines with size when females are larger. Rensch's rule, as this pattern has come to be known (Fairbairn 1997; Stillwell et al. 2010), has been found in some cases at the intraspecific level. However, among-population SSD typically does not conform to the rule (Blanckenhorn et al. 2006), and environmental conditions affect the degree rather than direction of SSD, often as a consequence of sex-related differences in growth and instar number (Teder and Tammaru 2005). Where Rensch's rule is found, it seems likely that greater plasticity in males than in females (the differential plasticity hypothesis), rather than sexual selection, might account for variation in SSD among populations (Fairbairn 2005). Sex-related differences in plasticity may also be the mechanism underlying a geographic version of Rensch's rule (Blanckenhorn et al. 2006). In addition to variation in phenotypic plasticity, strong sex-specific selection on size associated with season length may be responsible for the pattern (Blanckenhorn et al. 2006). Variation in nutritional resources and moisture may also play a role (Stillwell et al. 2007). Geographic variation in the relative sizes of males and females, its implications for SSD, and the mechanisms underlying this variation require further investigation (see Stillwell et al. 2007 and Hu et al. 2010, for examples; and Stillwell et al. 2010, for review).

In several species of insects strong relationships are found between the sizes of particular morphological traits (e.g., horns, mandibles) and body size, typically in only one of the sexes, such that large individuals have exaggerated morphologies, that is, a steep scaling relationship between body size and morphological trait size (Emlen and Nijhout 2000; Tomkins and Brown 2004). Both curvilinear and sigmoid allometries result in bimodal distributions of the traits in question (Emlen et al. 2005), and these dimorphisms are apparently underpinned by threshold mechanisms (Tomkins and Hazel 2011). However, matters may also be more complicated, involving two thresholds and facultative trimorphism in some beetle males (Rowland and Emlen 2009). The physiological underpinnings of these threshold characteristics are reasonably well understood (e.g., Emlen and Allen 2004; Emlen et al. 2007). It is clear that a wide variety of allometries exist, that in dung beetles horns can evolve and be lost rapidly, and that exaggerated traits can be extensively modulated in response to environmental conditions (Moczek and Nijhout 2003; Tomkins and Brown 2004; Emlen et al. 2007; Rowland and Emlen 2009). The ways in which different physical and social environments affect trait expression are the subject of considerable and ongoing research (e.g., Emlen 1997b, 1997a; Moczek and Nijhout 2003; Tomkins et al. 2005; Moczek 2006; Tomkins et al. 2006; Whitman and Ananthakrishnan 2009), with much emphasis on beetles, and particularly the scarabaeines (see reviews in Simmons and Ridsdill-Smith 2011).

The mechanisms underlying intraspecific body size variation clearly have to do with the determinants of size at maturity, and the way in which this size responds to both natural and sexual selection. The life history literature is replete with theoretical and empirical investigations of interactions and trade-offs between production rate, growth rate, age, mortality, fecundity, sex, mating strategies, season length, food quality, and temperature, and how these interact to influence and are often influenced by body size (reviewed by, among others, Kozłowski 1992; Stearns 1992; Blanckenhorn 2000; Roff 1992, 2001, 2002; Kozłowski et al. 2004; Blanckenhorn et al. 2006; Whitman 2008; Angilletta 2009; Whitman and Ananthakrishnan 2009). Although many aspects of these matters have been well explored, several important questions remain unresolved, of which perhaps the most significant in the context of this review are the mechanisms underlying large-scale clines in body size (see Roff 1980; Nylin and Gotthard 1998; Chown and Gaston 1999, 2010; Angilletta and Dunham 2003; Blanckenhorn and Demont 2004; Kozłowski et al. 2004; Angilletta 2009; Stillwell et al. 2010).

The intraspecific mechanisms underlying final adult size within species may also determine the interspecific body size frequency distribution. Kozłowski and Weiner (1997) present a detailed model showing how the mechanisms that underlie body size optimization at the intraspecific level can produce both interspecific species–body size distributions and interspecific allometries. Moreover, in a more recent model,

Kozłowski et al. (2003) proposed that genome size determines cell size, which in turn has an effect on metabolic rate and the way in which body size is optimized (via increases in cell size or number). Thus, body size and metabolic rates covary as a consequence of similar mechanisms and the influence of natural selection on genome size. These models make very different predictions regarding the intra- and interspecific scaling of metabolic rate to the nutrient supply network model developed by West et al. (1997; see Brown et al. 2004, for a review). For insects these alternative predictions are only beginning to be explored (see Chown et al. 2007; Irlich et al. 2009) in the context of the several models that exist to explain the scaling of metabolic rates and their thermal dependence (e.g., Brown et al. 2004; Glazier 2005, 2010; Kooijman 2010).

While there is substantial feedback between body size and life history variables (see Chown and Gaston 2010), the influence of intraspecific body size variation is nevertheless pervasive. Body size is related to mortality from abiotic factors such as starvation and desiccation (Lighton et al. 1994; Prange and Pinshow 1994; Ernsting and Isaaks 1997; Arnett and Gotelli 2003), mating success (Benjamin and Bradshaw 1994; Stone et al. 1995), fecundity (Juliano 1985; Honěk 1993; Tammaru et al. 1996), predation intensity and guild composition (Gaston et al. 1997; Nylin and Gotthard 1998), activity and foraging time (Stone 1994; Cerdá and Retana 2000), the outcome of intraspecific competition (Heinrich and Bartholomew 1979), flight ability (DeVries and Dudley 1990; Marden 1995; Roberts et al. 2004), and, of course, various aspects of morphology and physiology (e.g., Feener et al. 1988; Fairbairn 1992; Green 1999; Chown et al. 2007; Shingleton et al. 2008; Snelling et al. 2011). The nemopterid *Palmipenna aeoleoptera* has capitalized on the relationship between body size and predation likelihood by developing hypertrophied hind wings which deter small robberfly predators by creating an illusion of size (Picker et al. 1991).

INTERSPECIFIC. Data for insects have been employed in several of the classical studies of species–body size distributions (e.g., Boycott 1919; Hemmingsen 1934; Hutchinson and MacArthur 1959; May 1978). Nonetheless, understanding of such distributions for insects has been severely constrained by biased sampling, with larger species better represented than smaller ones. This said, where they have been reasonably well documented, species–body size distributions for major insect groups do not appear to differ markedly from those for other higher taxa. They are strongly right skewed on untransformed axes and typically approximately symmetric or right skewed when body size is logarithmically transformed, with departures from symmetry tending to become statistically more significant with increased numbers of species (e.g., May 1978; Hanski and Cambefort 1991; Barlow 1994; Dixon et al. 1995; Novotný and Kindlmann 1996; Loder 1997; Loder et al. 1997; Novotný and Wilson 1997; Brändle et al. 2000; Dixon and Hemptinne 2001; Maurer 2003; Ulrich 2004; Agosta and Janzen 2005; Finlay et al. 2006). Most insect species are small, but the smallest species are not the most frequent. The greater numbers of small-bodied species does not, however, translate into simple negative relationships in insects between the species richness and body size of taxonomic groups (e.g., Katzourakis et al. 2001; Orme et al. 2002a; Finlay et al. 2006; as has also been found in some more general studies—Orme et al. 2002b).

At lower levels of the taxonomic hierarchy, animal taxa tend to exhibit quite variable species–body size distributions (Maurer 1998; Gardezi and da Silva 1999; Kozłowski and Gawelczyk 2002). Thus, while overall the British beetles show a markedly right skewed distribution when body length is logarithmically transformed, different families exhibit negative, nonsignificant, and positive skew (Loder 1997). Indeed, the distribution of skewness values for those families with more than fifty species is itself peaked, but with most families being positively skewed. By contrast, overall for the insects, size frequency distributions are similar at a variety of spatial scales, with four modes, representing (1) Scolytidae and Chironomidae; (2) Curculionidae, Staphylinidae, and Chrysomelidae; (3) Noctuidae; and (4) Nymphalidae (Finlay et al. 2006).

The lognormal distribution has traditionally been regarded as an appropriate null model against which to test the form of species–body size distributions (e.g., Maurer et al. 1992; Dixon et al. 1995; Novotný and Kindlmann 1996; Gómez and Espadaler 2000; Espadaler and Gómez 2002). A common argument has been that, following from the central limit theorem, a variable subject to a moderately large number of independent multiplicative effects will tend to be lognormally distributed, and that body size can be thought of as just such a variable because growth is a multiplicative process. Unfortunately, the reasoning is wrong. The problem is directly analogous to that identified by Pielou (1975) with respect to the application of the same argument to species abundance distributions (see Williamson and Gaston 2005, for a recent discussion). The body sizes of individuals of any one species may be random variates from a lognormal distribution, provided that interactions between the individuals do not markedly influence their body size and that they have identical growth parameters, although as we have seen, in practice intraspecific body size distributions seem on the present limited evidence seldom to be lognormal or approximately lognormal. Even were they to be, it does not follow, however, that the distribution of body sizes of a number of different species occurring in an area must also be lognormal. Only if the species are separate independent samples of the same entity, with the same growth parameters, will the central limit theorem hold. Given the different conditions under which different species evolve and develop, this is an unlikely scenario.

Other mechanistic models, not always mutually exclusive, for the shape of species–body size distributions in general are based on (1) the distributions of optimal sizes resulting from an interspecific trade-off between production and mortality (as highlighted in the previous section; Kozłowski and Weiner 1997; Kindlmann et al. 1999); (2) patterns of speciation and extinction rates (Dial and Marzluff 1988; McKinney 1990; Maurer et al. 1992; Johst and Brandl 1997); (3) the world being larger, or the environment or resources being more finely subdivided, for smaller species (Hutchinson and MacArthur 1959; May 1978); and (4) patterns of dispersal (Chown 1997; Etienne and Olff 2004).

For insects, the importance of the size structure of resources has perhaps attracted the bulk of explicit attention, for herbivores (influenced by plant size structure; Dixon et al. 1995; Novotný and Wilson 1997), predators (influenced by prey size structure; Dixon and Hemptinne 2001), and parasites (influenced by host size—sometimes known as Harrison's rule; Tompkins and Clayton 1999; Johnson et al. 2005). Indeed, it appears that the size structure of the environment must have a profound influence on insect species–body size distributions (Morse et al. 1985). This is not at odds with species–body size distributions being shaped by intraspecific trade-offs between production and mortality, because resources affect the production function (itself the difference between assimilation and respiration; Kozłowski and Gawelczyk 2002).

Insects exhibit a wide range of dispersal abilities, from small species experiencing uncontrolled dispersal in the aerial plankton, through species with directed flight potentially of long duration, to species with limited dispersal abilities through flightlessness. This means that discussions of the role of dispersal in shaping species–body size distributions, predominantly concerning microbes, may be equally relevant to insects (e.g., Finlay 2002; Fenchel and Finlay 2004; Finlay et al. 2006). Here, the argument is that the smallest species are less speciose because they are widely distributed and disperse well, and thus the likelihood of speciation by isolation is depressed. The largest species are also less speciose, because they are such poor dispersers that allopatric speciation is depressed. Intermediate-sized species fall between these two constraints on speciation.

In a related vein, both at local and regional scales, the influence on species–body size distributions of transient or tourist species—those species present in an assemblage whose individuals obtain little if any of their nutrition directly or indirectly from resource bases that are present (Gaston et al. 1993)—has been a recurrent concern (e.g., Chown and Steenkamp 1996). Likewise, at local scales, in insect assemblages the densities of species have typically been found at best to be weakly negatively related, and perhaps more frequently unrelated, to their body sizes (e.g., Morse et al. 1988; Basset and Kitching 1991; Gaston et al. 1993; Basset 1997; Siemann et al. 1999b; Krüger and McGavin 2000). Evidence that this pattern generalizes to greater spatial extents, let alone global scales, is scant (local studies alone may involve the identification of tens of thousands of individuals) (see also a discussion in Diniz-Filho et al. 2010). However, those studies that have been conducted over greater extents provide little support for the notion that there is any simple relationship between abundance and body size in insects (e.g., Gaston 1988b; Gaston and Lawton 1988; Gutiérrez and Menéndez 1997). Abundance–body size relationships in birds have also proven generally to be rather weak (Gaston and Blackburn 2000), leading to the suggestion that the existence or otherwise of such relationships may depend heavily on the dispersal characteristics of a taxonomic group.

## Variation through Time

### *Evolutionary Trends*

Current evidence suggests that the insects arose from a common ancestor at the Silurian-Ordovician boundary (ca. 420 Myr BP; Grimaldi and Engel 2005). The fossil record for early insects and closely related groups is, however, poor. Winged insects first appear in the fossil record in the Carboniferous (ca. 325 Myr BP), a period when the group began to radiate (Grimaldi and Engel 2005). Although diversification was disrupted by several extinction events (see, e.g., Wilf et al. 2006), the diversity of the group has continued to rise.

A much-discussed phenomenon among the insects is the gigantism that was taxonomically widespread in the late Paleozoic, including among the Protodonata, Paleodictyoptera, Ephemeroptera, Diplura, and Thysanura (Kukalová-Peck 1985, 1987; Shear and Kukalová-Peck 1990; Graham et al. 1995; Dudley 1998; Wootton and Kukalová-Peck 2000). One vigorously championed mechanism for the occurrence of gigantism during this period (which also occurred in other invertebrate and lower vertebrate groups) is hyperoxia and hyperbaria in the Paleozoic atmosphere, leading to a relaxation of constraints on tracheal diffusion and the power demands of flight musculature in winged species (Miller 1966; Graham et al. 1995; Dudley 1998, 2000b, 2000a; Berner et al. 2003; Harrison et al. 2010). Oxygen availability would also have been enhanced in the aquatic larval stages of many of the groups, though gigantism was just as common in terrestrial species. This oxygen pulse hypothesis is partially consistent with the subsequent loss of these forms with increasing hypoxia in the late Permian (Huey and Ward 2005), and the evolution of large size in at least one group (the mayfly family Hexagenitidae) during a secondary oxygen peak in the Cretaceous (Dudley 2000a). Although appealing, the atmospheric partial pressure hypothesis (more conveniently the "a$PO_2$ hypothesis"—see Harrison et al. 2010) has typically not been well explored within a strong inference framework.

However, the situation is now changing rapidly, and several recent studies have not only provided new empirical data concerning the physiological basis of interactions between atmospheric (or dissolved) oxygen partial pressure and size (e.g., Peck and Maddrell 2005; Klok and Harrison 2009; Klok et al. 2009; Klok et al. 2010; Verberk et al. 2011), but have also provided a rigorous theoretical framework within which to interpret new information (Harrison et al. 2010). Empirical work has shown that performance declines in hypoxia but is less affected by hyperoxia, and that critical $PO_2$ levels are generally size invariant (Harrison et al. 2010). However, critical $PO_2$ values do rise in *Manduca sexta* late in each larval stage as tissue demand increases, but the tracheal system does not increase in size (Greenlee and Harrison 2005). Such changes have pronounced effects on the processes of molting and may act as a size-dependent signal for these (Callier and Nijhout 2011). Hypoxia causes molting at a smaller size, while hyperoxia results in molting at a larger size, though the effect is less than that of hypoxia. Rearing in variable oxygen partial pressures also results in a change in size in many (but not all) other insect species, with size declining markedly in hypoxia but increasing to a lesser extent in hyperoxia (reviewed in Harrison et al. 2010). Over several generations in *Drosophila* the decline in size with hypoxia seems to owe exclusively to plasticity, because sizes typical of ambient conditions are restored after reexposure to these conditions (Klok et al. 2009). By contrast, size increases after several generations in hyperoxia are maintained and may in part be due to maternal effects.

Among the most promising work in this area is the suggestion that limitations to size are set by steeper scaling of the cross-sectional area of the tracheae in the legs ($mass^{1.02}$) than of the cross-sectional area of the leg orifice ($mass^{0.77}$) (Kaiser et al. 2007). These scaling differences mean that space available for tracheae in the legs may limit the size of an insect, and such limitations may well apply to other constrictions of the exoskeleton. Under experimental, hyperoxic conditions, insects reduce the dimensions of their tracheae, which suggests that during such conditions in the past, the constraints on large size may have been reduced substantially, so leading to gigantism (Kaiser et al. 2007). However, empirical work in the area is limited and requires additional exploration for a range of taxa (Harrison et al. 2010).

Several alternative mechanisms, some of which involve oxygen partial pressure, might also explain gigantism and its disappearance in the more modern fauna. Staying with oxygen partial pressure, the evolution of large size and its subsequent decline might have involved interactions between oxygen availability, geographic range size, and thermal tolerances. Reductions in geographic range size are thought to have been precipitated both by warming and by a decline in atmospheric oxygen (Huey and Ward 2005), and typically larger species require larger geographic ranges to avoid extinction. Aquatic stages may also have been compromised more than their terrestrial counterparts (see the review of size, partial pressure, and solubility patterns by Verberk et al. 2011) owing to differential changes in metabolic demand and oxygen flux (Huey and Ward 2005; Makarieva et al. 2005), although flight metabolism in modern insects has a much lower critical $PO_2$ than resting metabolism (Rascón and Harrison 2005). Hypoxic stress may also have lowered thermal tolerance (Pörtner 2001), so reducing range sizes, especially if upper and lower thermal limits are decoupled as they appear to be in insects. Moreover, in some groups, other aspects of performance may have been affected (Verberk et al. 2011). However, the evidence for oxygen limitation of thermal tolerance in insects remains equivocal (Chown and Terblanche 2007; Stevens et al. 2010).

Other hypotheses, which do not involve oxygen partial pressure variation (see also Harrison et al. 2010), include the evolution of large size as a defensive adaptation of Paleozoic arthropods directed toward predation by vertebrates, of which the majority at the time were insectivorous or predators on other vertebrates (Shear and Kukalová-Peck 1990). Changes in size, particularly later reductions, might also have been mediated by changing mortality risks that must have been encountered by juvenile stages. In this context it is notable that the largest recent insects (extant or recently extinct) either typically spend the bulk of their lives as concealed feeders (e.g., beetle species in the Cerambycidae, Scarabaeidae, Dynastinae) or are restricted to oceanic islands where predation pressure may be lower (e.g., St. Helena giant earwig *Labidura herculeana*, New Zealand giant weta *Deinacrida* spp.). In addition, the late Permian not only saw the loss of giant insects, but a mass extinction of insect diversity (Labandeira and Sepkoski 1993). Assuming that, relative to the absolute giants, most insect species were nonetheless relatively small bodied before this event, and that in keeping with virtually all species–body size distributions that have ever been documented the largest-bodied species were relatively species poor, then even random losses of species with respect to body size would almost certainly have seen the loss of the more giant forms.

Which of these mechanisms is likely to have had the predominant role in promoting gigantism, and its subsequent disappearance, is difficult to determine, but the question deserves further exploration in the context of the factors determining final body size in insects.

The evolution and subsequent disappearance of several comparatively large-bodied taxa in the insects arguably draws attention away from a somewhat different evolutionary trend in size found in other groups of organisms—that is, the tendency for selection acting on individual organisms predominantly to favor larger body size (including insects), resulting, if unopposed, in a macroevolutionary trend toward increased size (Kingsolver and Pfennig 2004). Such a macroevolutionary pattern is known as Cope's rule (Benton 2002). Clearly the existence of many large-bodied forms early in the evolution of the insects means that this rule has not been obeyed over the entire duration of the insects, either for the group as a whole, or for several major clades. However, the picture may appear rather different if one focuses on the period after the

Permian mass extinction. Endopterygote insects predominate in recent insect faunas, particularly those in the orders Coleoptera, Hymenoptera, Diptera, and Lepidoptera (Gaston 1991). Although the ancestors of at least some of these groups were present in the Permian, they all underwent dramatic and continued diversification after the mass extinction event and have continued to do so through to the present (Labandeira and Sepkoski 1993; Gaunt and Miles 2002; Grimaldi and Engel 2005). In all four orders it seems likely that the largest recent species are among, if not actually, the largest that have existed (Chown and Gaston 2010). Within particular clades, phyletic size increase has been explored for only a single family, the carabids. Of the thirty-four groups examined, seven showed significantly positive correlations between body size and cladogram position (indicating phyletic size increase), two showed significantly negative relationships, and in the remainder there was no relationship between size and cladogram position (Liebherr 1988). Although a macroevolutionary trend toward large size is thus uncommon in the family, it is not randomly distributed among taxa. Typically, phyletic size increase is associated with brachyptery, and with groups inhabiting stable environments, although the mechanisms responsible for this pattern have not been fully explored.

Both modern experimental and field ecology and palaeontology reveal how rapidly body size changes can evolve and how developmentally plastic it can be (see also Frankino et al. 2005). In the former case, David et al. (1997) demonstrated that laboratory-reared *Drosophila melanogaster* show both a reduction in size and change in kurtosis of the size frequency distribution relative to their wild counterparts. In the latter, numerous experiments have shown that size changes can be effected rapidly within generations depending on external conditions such as food availability or oxygen tension (e.g., Emlen and Nijhout 2000; Frazier et al. 2001; Klok et al. 2009), and that selection can effect rapid size changes between generations (e.g., Gibbs et al. 1997). In the field, rapid evolution of body size has been shown in *Drosophila subobscura* (Huey et al. 2000). This species is native to the Old World, where it displays a positive cline in wing length with latitude. It was, however, introduced to North and South America, spreading rapidly and evolving a cline in body size that largely converged on that observed in the Old World (although the way in which the variation in size was achieved is different). Likewise, the threshold for horn development in the beetle *Onthophagus taurus* evolved rapidly, in opposite directions, in populations introduced to western Australia and eastern North America from the Mediterranean region (Moczek and Nijhout 2003). Much emphasis is now being placed on understanding the quantitative genetic basis of size variation and how responses to the environment are mediated through plasticity, selection, and evolutionary changes in plasticity (see, e.g., Kennington et al. 2007; Stillwell and Fox 2009; Kennington and Hoffmann 2010; Lee et al. 2011).

### *Ecological Trends*

INTRASPECIFIC. At the intraspecific level, body size has been shown to vary in several groups of insects both within and between years. Within a season, final size is strongly dependent on interactions between time constraints, resource allocation to growth and/or reproduction, mortality, aging, and food quality (Kozłowski 1992; Ayres and Scriber 1994; Abrams et al. 1996; Nylin and Gotthard 1998; Taylor et al. 1998; Kozłowski and Teriokhin 1999; Cichoń and Kozłowski 2000; Scriber 2002; De Block and Stoks 2008; Whitman 2008). In several species, despite initially poor resource (including time and temperature) conditions, there is elevation of growth rate such that final body sizes of adults show much less variation than might otherwise have been the case (Nylin and Gotthard 1998; Gotthard et al. 2000; Margraf et al. 2003; Tseng 2003; Strobbe and Stoks 2004). Nonetheless, additional growing time does not always serve to increase size (see Kause et al. 2001; Berner et al. 2005). Compensatory (or catch-up) growth is associated with intrinsic physiological costs, such as developmental errors and oxidative damage, and extrinsic costs such as increased exposure to predators and/or parasitoids (Nylin et al. 1996; Mangel and Munch 2005; Stoks et al. 2006a; De Block and Stoks 2008). Thus, resource quality and the presence of predators often play a significant role in determining final size, whether this size is attainable given seasonal time constraints, and what the costs thereof might be (Scriber 2002; Berner et al. 2005; Stoks et al. 2006b; De Block and Stoks 2008; Röder et al. 2008). Although increased growth rates might be able to compensate for linear dimensions, mass may nonetheless be reduced under time constraints (Nylin and Gotthard 1998; Strobbe and Stoks 2004). At least in income breeders, adults can gain mass by feeding, reducing the significance of low emergence mass. For capital breeders, emergence mass is set by resource allocation in the immature stages. This difference between income and capital breeders affects the relationship between final size and size variation because in the latter a linear increase in fecundity with size is likely, while in the former behavioral performance constrains maximum size (Teder et al. 2008). Although many adult insects do not live for more than a season, there are species in which adults may be long lived (such as ant queens, some beetles, and some butterflies; see Rockstein and Miquel 1973). For species such as these, resource allocation models and the relationships between age and size at reproduction have been well explored (see Kozłowski and Wiegert 1987).

Seasonal variation in body size, in species where there is no period of diapause or quiescence, has been described for a wide variety of species, including mosquitoes (Yuval et al. 1993), blackflies (fig. 1.2; Colbo and Porter 1979; Baba 1992; Myburgh 2001), *Drosophila* (Tantawy 1964; Kari and Huey 2000), tsetse flies (Rogers and Randolph 1991), beetles (Ernsting and Isaaks 1997), stoneflies (Haro et al. 1994), a parasitoid wasp (Sequeira and Mackauer 1993), a bee (Alcock et al. 2006), and butterflies (Rodrigues and Moreira 2004). In the majority of these cases developmental temperature has the most significant influence on body size, such that size tends to be largest at the lowest temperatures (see Atkinson 1994), although resource availability or stressful abiotic conditions may also play a role (Baba 1992; Kari and Huey 2000), which is in keeping with many investigations of resource competition and the effects of stress on insects (see discussion in Whitman 2008; Chown and Gaston 2010).

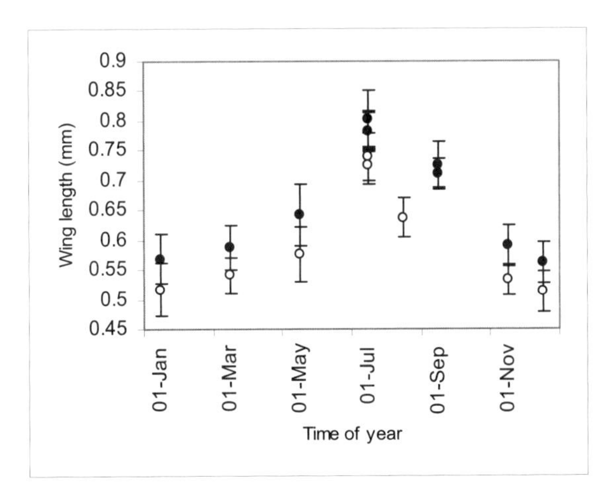


FIGURE 1.2. Seasonal variation in mean (± SE) wing length (mm) of female (*solid symbols*) and male (*open symbols*) *Simulium chutteri* collected over the course of a single year along the Orange River in South Africa (data from Myburgh 2001).

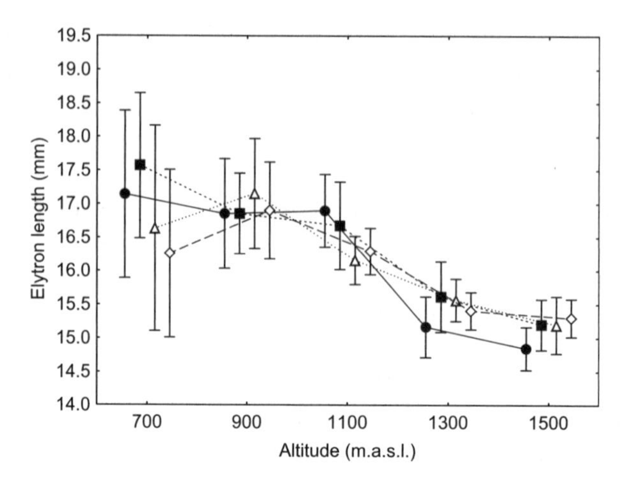


FIGURE 1.3. Elytron length variation (mean ± 95% C.I.) along an altitudinal gradient in the Cederberg district of South Africa for *Thermophilum decemguttatum* males. Four years are shown (*solid line and solid circles*: October 2002; *short-dashed line and solid squares*: October 2003; *long-dashed line and open triangles*: October 2004; *dotted line and open diamonds*: October 2005). Redrawn from Chown and Gaston (2010).

Interannual variation in size has also been investigated in several species (Alcock 1984; Evans 2000; Smith et al. 2000). Such variation is usually not substantial (Evans 2000; Smith et al. 2000), and the likely causes of variation have not been systematically explored (Chown and Gaston 2010). In analyses of altitudinal and interannual variation in the elytron length of the carabid *Thermophilum decemguttatum* (fig. 1.3) that first included only year and altitude as factors and then included site mean temperature and growing season length, Gouws (2007) showed that temperature and growing season length exclude the year factor, suggesting that a combination of temperature and time available for development underlies body length variation between years. The ultimate mechanisms determining this relationship may well be similar to those responsible for spatial variation in size (see below) and must be mediated by a variety of proximate physiological mechanisms (see Davidowitz et al. 2005; and discussion in Chown and Gaston 2010).

INTERSPECIFIC. Both at individual sites through time as succession progresses and across sites of different successional status, the average body sizes of species in insect assemblages tend to decline with succession (e.g., Steffan-Dewenter and Tscharntke 1997; Siemann et al. 1999a;

Braun et al. 2004). This is despite the species richness of different groups increasing, decreasing, and remaining approximately stable. A similar trend in body size tends to occur along gradients of increasing disturbance (e.g., Blake et al. 1994; Grandchamp et al. 2000; Ribera et al. 2001; Braun et al. 2004). Both patterns probably result from changes in the environmental constraints on body size, particularly those associated with vegetational complexity and stability. Similar changes may be found as a consequence of changing vegetation structure associated with biological invasions by plant species, though less evidence exists for size changes in insects associated with direct human impacts on habitats (Chown and Gaston 2010). The invasion of islands by insects might well be size dependent, thus having a significant influence on size distributions, although the mechanisms underlying such patterns are poorly understood. Lawton and Brown (1986) reported a negative relationship between the probability of establishment of an invader accidentally or intentionally introduced to the British Isles and its body size. A similar pattern was found within higher taxa for the insects of sub-Antarctic Marion Island (Gaston et al. 2001).

## Variation through Space

Perhaps one of the most controversial areas in the macroecological literature is the question of spatial variation in body size: what this variation should be termed (i.e., whether it should be formally named as one or more rules), how it arises, and how it should be investigated. Although some of these matters have been considered explicitly at least since Bergmann's late nineteenth-century work (see reviews by Blackburn et al. 1999; Watt et al. 2010), no signs of a flagging interest therein by biologists can be discerned. Indeed, the field is as vigorous today as it seems ever to have been (see, e.g., Watt et al. 2010; Huston and Wolverton 2011; Olalla-Tárraga 2011; Watt and Salewski 2011). At least part of the explanation for the current situation lies with the ecological and physiological significance of body size (e.g., Brown et al. 2004), indications that substantial size changes may be taking place in response to anthropogenic climate change (see discussions in Gardner et al. 2011; Yom-Tov and Geffen 2011), and the abundant scope that Bergmann's rule seems to offer for philosophical reflections on generalities in ecology. The form of geographic variation in body size and the mechanisms underlying it in insects have played a significant role in these discussions (e.g., Arnett and Gotelli 2003; Chown and Klok 2003). In this section, we therefore provide a brief overview of geographic patterns in insect size variation and the mechanistic hypotheses that have been proposed to explain them (a comprehensive recent treatment is provided by Chown and Gaston 2010). Nonetheless, before doing so we should draw attention to the fact that the large majority of insects, and indeed of most organisms, have relatively small range sizes. Therefore, spatial variation in size, at least over relatively large geographic distances, perhaps applies only to a relatively small portion of any regional assemblage (Gaston et al. 2008). Local spatial size variation is an entirely different matter (e.g., Warren et al. 2006) but is not our concern here.

### *Intraspecific Patterns*

The two primary patterns of spatial variation in insect body size are those concerned with latitudinal and altitudinal gradients (fig. 1.4). In many species, body size (usually measured as a characteristic linear dimension) increases with latitude and/or with altitude. Although the literature in this field is growing all the time (e.g., Cassell-Lundhagen et al. 2010), latitudinal and altitudinal size increases have been found in more than twenty insect species, including members of the Coleoptera, Diptera, Hemiptera, Hymenoptera, Neuroptera, and Orthoptera (Chown and Gaston 2010). In many instances it has been shown that these patterns are not simply a consequence of plasticity, because size variation is retained when populations are grown under common garden conditions (James et al. 1995; Arnett and Gotelli 1999; Huey et al. 2000; Loeschcke et al. 2000; Weeks et al. 2002; Gilchrist and Huey 2004; Hu et al. 2011; see also Cassell-Lundhagen et al. 2010). Size declines with latitude or altitude have been found in almost as many species, including those in the orders Coleoptera, Diptera, Lepidoptera, Neuroptera, and Orthoptera. In some of these species, the declines are not constant but take the form of a sawtooth cline, such that increasing season length leads to increasing body size until two generations can be incorporated within a season, at which point body size declines precipitously (Roff 1980; Masaki 1996; Nygren et al. 2008).

The mechanisms underlying intraspecific size clines in insects and other ecototherms have been much debated (see, e.g., Nylin and Gotthard 1998; Chown and Gaston 1999; Blanckenhorn and Demont 2004;

(a)

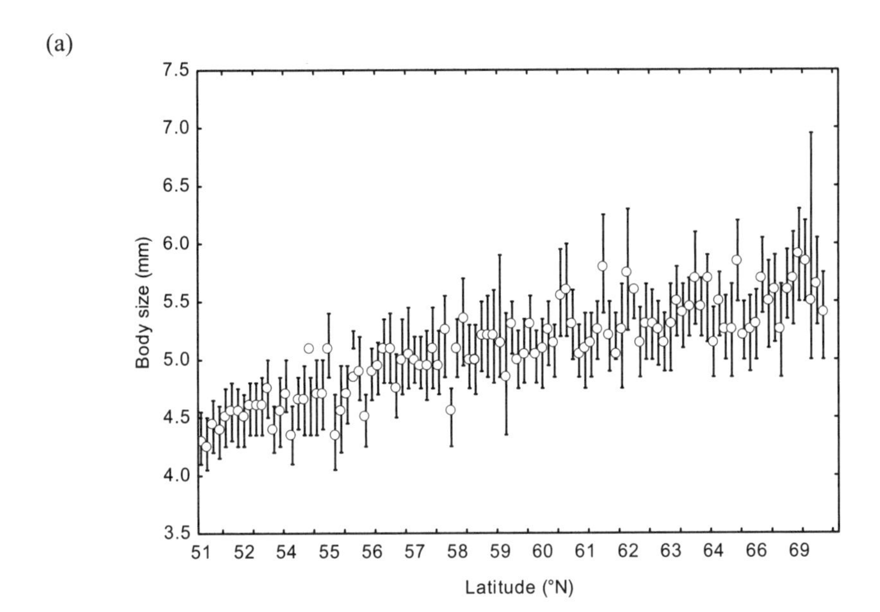

Body size (mm)
7.5
7.0
6.5
6.0
5.5
5.0
4.5
4.0
3.5
51 52 54 55 56 57 58 59 60 61 62 63 64 66 69
Latitude (°N)


(b)

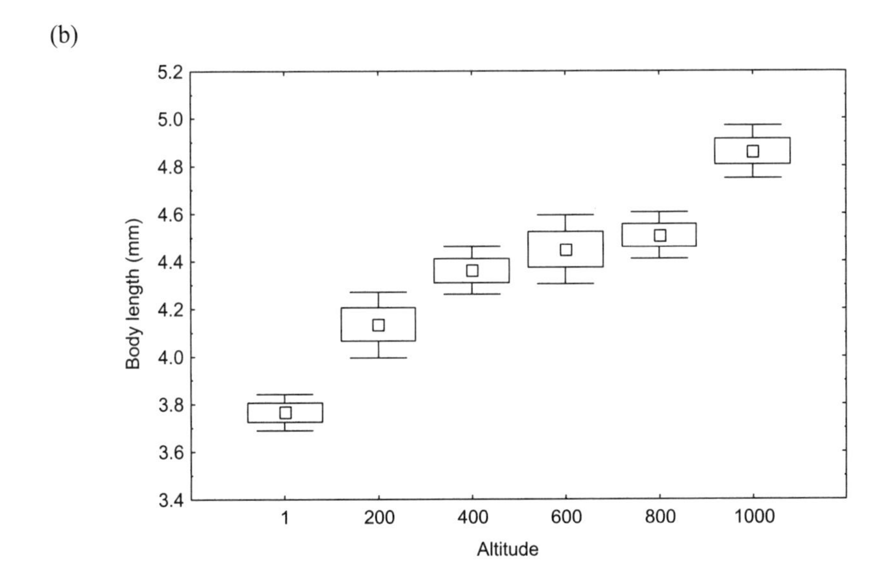

Body length (mm)
5.2
5.0
4.8
4.6
4.4
4.2
4.0
3.8
3.6
3.4
1 200 400 600 800 1000
Altitude

(c)

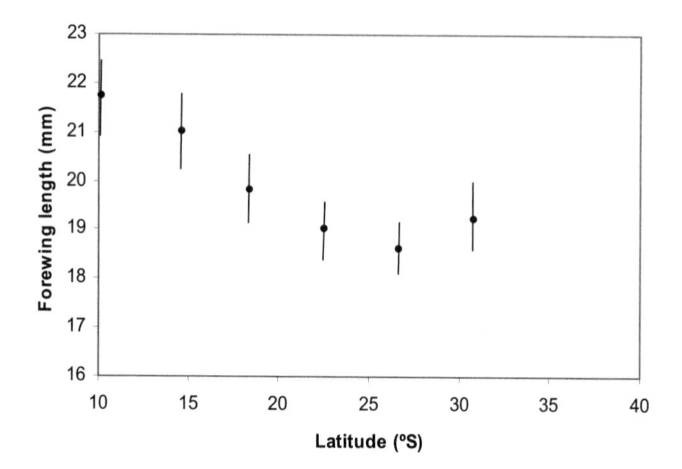

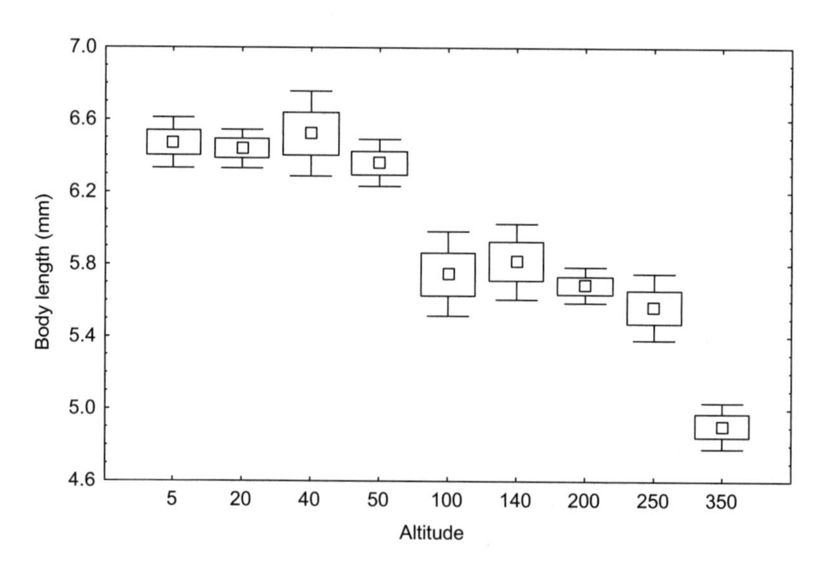


FIGURE 1.4. Geographic patterns in species body size include (*a*) latitudinal increases, as in ant species in the British Isles and northern Europe; (*b*) altitudinal increases, as in the weevil *Bothrometopus parvulus* from Marion Island (mean, SE, and 95% CI); (*c*) latitudinal declines, as in butterfly species in Australia; and (*d*) altitudinal declines, as shown here for the weevil *Ectemnorhinus viridis* from Heard Island (mean, SE, and 95% CI). (*a*) redrawn from Cushman et al. (1993), (*b*) and (*d*) from Chown and Klok (2003), and (*c*) from Hawkins and Lawton (1995).

Gotthard 2004; Kozłowski et al. 2004; Angilletta 2009). At least eight mechanistic hypotheses which can be applied to insects have been proposed and suggestions made that the existing evidence for insects might support them. These include the proximate biophysical model (van der Have and de Jong 1996; de Jong and van der Have 2009), temperature threshold hypothesis (Walters and Hassall 2006), optimal resource allocation model (Kozłowski et al. 2004), minimum metabolic rate hypothesis (Makarieva et al. 2005), starvation resistance hypothesis (Cushman et al. 1993), resource availability hypothesis (reviewed most recently by Huston and Wolverton 2011), heat balance hypothesis (Olalla-Tárraga and Rodríguez 2007), and combined resource allocation model (Chown and Gaston 1999, 2010). The majority have been reviewed by Chown and Gaston (2010), who also set out the alternative predictions of these hypotheses in terms of their various expectations, such as the direction of the clines (negative, positive, or sawtooth), the significance of larval mortality factors, and differences in the scaling of rates and storage components (such as the scaling of lipid quantity vs. metabolic rate). Much scope for examination of these hypotheses in a strong inference context still exists, given that many previous studies tended to focus on a limited subset thereof. However, evidence is accumulating that the combined resource allocation model, which combines Roff's (1980) model explaining sawtooth clines and Kozłowski et al.'s (2004) resource allocation models, may provide the most comprehensive explanation for all forms of intraspecific clines in size. Not only can they account for the variation in clines, but they also explain the fact that smaller species tend to show positive increases with latitude and/or altitude, while larger species tend to show the converse (Chown and Klok 2003; Blanckenhorn and Demont 2004). Moreover, they can also be tied readily to the proximate physiological mechanisms which underlie variation in size (Davidowitz et al. 2005; Nijhout et al. 2010).

Despite clear evidence that latitudinal and altitudinal variation in body size in either direction could be adaptive, much remains to be done to elucidate whether the mechanisms proposed to underlie this adaptive variation enjoy empirical support. Substantial complexities underlie changes in body size (such as varying cell size and/or number in *Drosophila*; see Partridge et al. 1994; James et al. 1995; Partridge and Coyne 1997), and the repeated evolution of similar size gradients might take place in different ways (Huey et al. 2000). Moreover, careful investigations of interacting factors, such as clines in mortality, season length, temperature, and resource availability, are not common. If the models underlying adaptive variation in size are to be tested, there has to be an increase both in the depth (i.e., simultaneous investigation of competing explanations) and number of studies, and a recognition that multiple factors might concurrently affect size variation in ways that may be associated with the form of landscape structure and gene flow across the landscape.

### *Interspecific Patterns*

Interspecific body size clines in insects have also been widely examined. Such studies usually take one of two forms, distinguished by Gaston et al. (2008) as interspecific and assemblage analyses. Interspecific analyses treat different species as separate data points, typically plotting average size against the position (usually the midpoint) of the range of that species on a latitudinal or altitudinal gradient. In assemblage analyses the data points are different areas (sites), and typically the average size of the species occurring in an area is plotted against the position of that area on a latitudinal or altitudinal gradient. While related, the former captures how between-species variation in body size changes with the distribution of those species across the landscape, and the latter captures how variation in the body size composition of species assemblages changes across the landscape. For assemblage analyses, increases (Hawkins and Lawton 1995), declines (Barlow 1994), and no simple change (Hawkins 1995) in size with latitude have been documented, while for interspecific analyses, increases (Moreteau et al. 2003) and declines (Diniz-Filho and Fowler 1998) with latitude have been found. It has also been suggested that colony size of a given ant (or termite) species can be used as a measure of body size (of the superorganism) (Kaspari and Vargo 1995; Porter and Hawkins 2001) (see the reviews in Dillon et al. 2006; Chown and Gaston 2010).

Several adaptive arguments have been proposed for spatial change in average body sizes, with the most common hypothesis being enhanced tolerance of starvation or desiccation (Cushman et al. 1993; Kaspari and Vargo 1995; Blackburn et al. 1999; Chown and Gaston 1999). However, interspecific body size clines, especially if expressed as means for a given geographic location (i.e., assemblage-level analyses), are much more difficult to interpret than intraspecific geographic size variation. Clinal size variation in assemblage body size is a consequence of geographic changes in the location and shape of the interspecific species–body size distribution (Janzen et al. 1976 provide an early example). Thus, the form of assemblage size clines depends on species richness (see Greve et al. 2008), beta diversity (the spatial pattern of species gains and losses—see Koleff et al. 2003), and the form of the intraspecific size clines of the species that are retained across more than two sites. Therefore, clines could take virtually any form. The latter has been demonstrated several times, as has the influence of spatial turnover of higher taxa on the form of the interspecific size cline (Hawkins and Lawton 1995; Chown and Klok 2003; Brehm and Fiedler 2004). Moreover, adaptive explanations at the assemblage level make the implicit assumption either that the average body size of the assemblage is being optimized or that a certain size is optimal for a given reason, which seems unlikely (Chown and Gaston 2010).

Is there, therefore, any merit in investigating assemblage-level spatial variation in size? Clearly, the answer depends on the context in which this is to be done. If the variation is characterized only by the mean body size of the assemblage, then investigations are perhaps not as useful as they could be. By contrast, the effects of spatial scale and spatial position on the central tendency and form of variation in assemblages, and what such effects mean for the mechanisms structuring assemblages, are clearly significant macroecological questions (Maurer 1998) that could benefit from exploration using data for insects.

## In Conclusion

There is evidence for a rich spatial and temporal structuring and variation in insect body sizes. However, many of the generalizations that can be made remain based on empirical investigations of a relatively narrow range of species, often drawn from a small number of higher taxa, as is the case for much of insect physiology (discussion in Chown et al. 2002; Chown and Nicolson 2004). The most significant question, then, is whether the patterns and mechanisms are expected to vary substantially between higher taxa and/or between species. At a superficial level, the answer to this question is obvious—there must be subtle differences between species. However, the more important question is whether there is a single set of underlying mechanisms that generate size variation through space and time that are not strongly contingent or subject to substantial phylogenetic constraint. At present there seems to be little agreement on an answer, which goes to the heart of the determinants of size variation and their consequences throughout the ecological hierarchy (Kozłowski et al. 2003; Brown et al. 2004). What is clear, however, is that in many instances tests of the proposed mechanisms must rely on a small, and often patchy, dataset that is poor with respect to the significance of insects in terms of global biodiversity (see also Frankino et al. 2005; Chown and Gaston 2010; Diniz-Filho et al. 2010). Remedying this situation should form a key goal if the evolution of size and its effects on biodiversity are to be fully comprehended.

## Acknowledgments

We are grateful to N. Loder and E. Myburgh for providing access to unpublished work, to T. M. Blackburn, J. Kozłowski, and an anonymous reviewer for comments on a previous version of the manuscript, and to P. Johnson and L. Knight for assistance. K.J.G. acknowledges the support of A. Mackenzie, and S.L.C. acknowledges support from the National Research Foundation. Material presented here includes an update of sections of work available in S. L. Chown and K. J. Gaston, "Body size variation in insects: A macroecological perspective," *Biological Reviews* 85 (2010): 139–169.

## References

Abrams, P. A. , O. Leimar, S. Nylin, and C. Wiklund. 1996. "The effect of flexible growth rates on optimal sizes and development times in a seasonal environment." *American Naturalist* 147:381–395.

Agosta, S. J., and D. H. Janzen. 2005. "Body size distributions of large Costa Rican dry forest moths and the underlying relationship between plant and pollinator morphology." *Oikos* 108:193.

Alcock, J. 1984. "Long-term maintenance of size variation in populations of *Centris pallida* (Hymenoptera: Anthophoridae)." *Evolution* 38:220–223.

Alcock, J., L. W. Simmons, and M. Beveridge. 2006. "Does variation in female body size affect nesting success in Dawson's burrowing bee, *Amegilla dawsoni* (Apidae: Anthophorini)?" *Ecological Entomology* 31:352–357.

Angilletta, M. J. 2009. *Thermal adaptation: A theoretical and empirical synthesis*. Oxford: Oxford University Press.

Angilletta, M. J., and A. E. Dunham. 2003. "The temperature-size rule in ectotherms: Simple evolutionary explanations may not be general." *American Naturalist* 16:332–342.

Arnett, A. E., and N. J. Gotelli. 1999. "Geographic variation in life-history traits of the ant lion, *Myrmeleon immaculatus*: Evolutionary implications of Bergmann's rule." *Evolution* 53 (4): 1180–1188.

——. 2003. "Bergmann's rule in larval ant lions: Testing the starvation resistance hypothesis." *Ecological Entomology* 28 (6): 645–650.

Atkinson, D. 1994. "Temperature and organism size—a biological law for ectotherms?" *Advances in Ecological Research* 25:1–58.

Ayres, M. P., and J. M. Scriber. 1994. "Local adaptation to regional climates in *Papilio canadensis* (Lepidoptera, Papilionidae)." *Ecological Monographs* 64 (4): 465–482.

Baba, M. 1992. "Oviposition habits of *Simulium kawamurae* (Diptera, Simuliidae), with reference to seasonal changes in body size and fecundity." *Journal of Medical Entomology* 29 (4): 603–610.

Barlow, N. D. 1994. "Size distributions of butterfly species and the effect of latitude on species sizes." *Oikos* 71 (2): 326–332.

Bartholomew, G. A., and T. M. Casey. 1978. "Oxygen consumption of moths during rest, pre-flight warm-up, and flight in relation to body size and wing morphology." *Journal of Experimental Biology* 76 (Oct): 11–25.

Basset, Y. 1997. "Species-abundance and body size relationships in insect herbivores associated with New Guinea forest trees, with particular reference to insect host-specificity." In *Canopy arthropods*, edited by J. A. Stork and R. K. Didham, 237–264. London: Chapman and Hall.

Basset, Y., and R. L. Kitching. 1991. "Species number, species abundance and body length of arboreal arthropods associated with an Australian rainforest tree." *Ecological Entomology* 16 (4): 391–402.

Benjamin, S. N., and W. E. Bradshaw. 1994. "Body size and flight activity effects on male reproductive success in the pitcherplant mosquito (Diptera, Culicidae)." *Annals of the Entomological Society of America* 87 (3): 331–336.

Benke, A. C., A. D. Huryn, L. A. Smock, and J. B. Wallace. 1999. "Length-mass relationships for freshwater macroinvertebrates in North America with particular reference to the southeastern United States." *Journal of the North American Benthological Society* 18:308–343.

Benton, M. J. 2002. "Cope's rule." In *Encyclopedia of evolution*, edited by M. Pagel, 209–210. Oxford: Oxford University Press.

Berner, D., W. U. Blanckenhorn, and C. Körner. 2005. "Grasshoppers cope with low host quality by compensatory feeding and food selection: N limitation challenged." *Oikos* 111:525–533.

Berner, R. A., D. J. Beerling, R. Dudley, J. M. Robinson, and R. A. Wildman. 2003. "Phanerozoic atmospheric oxygen." *Annual Review of Earth and Planetary Sciences* 31:105–134.

Billick, I., and C. Carter. 2007. "Testing the importance of the distribution of worker sizes to colony performance in the ant species *Formica obscuripes* Forel." *Insectes Sociaux* 54:113–117.

Blackburn, T. M., and K. J. Gaston. 1994. "The distribution of body sizes of the world's bird species." *Oikos* 70 (1): 127–130.

Blackburn, T. M., K. J. Gaston, and N. Loder. 1999. "Geographic gradients in body size: A clarification of Bergmann's rule." *Diversity and Distributions* 5:165–174.

Blake, S., G. N. Foster, M. D. Eyre, and M. L. Luff. 1994. "Effects of habitat type and grassland management practices on the body size distribution of carabid beetles." *Pedobiologia* 38 (6): 502–512.

Blanckenhorn, W. U. 2000. "The evolution of body size: What keeps organisms small?" *Quarterly Review of Biology* 75 (4): 385–407.

Blanckenhorn, W. U., and M. Demont. 2004. "Bergmann and converse Bergmann latitudinal clines in arthropods: Two ends of a continuum?" *Integrative and Comparative Biology* 44 (6): 413–424.

Blanckenhorn, W. U., R. C. Stillwell, K. A. Young, C. W. Fox, and K. G. Ashton. 2006. "When Rensch meets Bergmann: Does sexual size dimorphism change systematically with latitude?" *Evolution* 60:2004–2011.

Boycott, A. E. 1919. "On the size of things, or the importance of being rather small." In *Contributions to medical and biological research*, edited by C. L. Dana, 226–234. New York: Hoeber.

Brändle, M., J. Stadler, and R. Brandl. 2000. "Body size and host range in European Heteroptera." *Ecography* 23 (1): 139–147.

Braun, S. D., T. H. Jones, and J. Perner. 2004. "Shifting average body size during regeneration after pollution: A case study using ground beetle assemblages." *Ecological Entomology* 29 (5): 543–554.

Brehm, G., and K. Fiedler. 2004. "Bergmann's rule does not apply to geometrid moths along an elevational gradient in an Andean montane rain forest." *Global Ecology and Biogeography* 13 (1): 7–14.

Brown, J. H., J. F. Gillooly, A. P. Allen, V. M. Savage, and G. B. West. 2004. "Toward a metabolic theory of ecology." *Ecology* 85 (7): 1771–1789.

Callier, V., and H. F. Nijhout. 2011. "Control of body size by oxygen supply reveals size-dependent and size-independent mechanisms of molting and metamorphosis." *Proceedings of the National Academy of Sciences of the United States of America* 108:14664–14669.

Cassell-Lundhagen, A., P. Kaňuch, M. Low, and Å. Berggren. 2010. "Limited gene flow may enhance adaptation to local optima in isolated populations of Roesel's bush cricket (*Metroptera roeselii*)." *Journal of Evolutionary Biology* 24:381–390.

Cerdá, X., and J. Retana. 2000. "Alternative strategies by thermophilic ants to cope with extreme heat: Individual versus colony level traits." *Oikos* 89 (1): 155–163.

Chapman, A. D. 2009. *Numbers of living species in Australia and the world*. 2nd ed. Canberra: Australian Biological Resources Study (ABRS).

Chown, S. L. 1997. "Speciation and rarity: Separating cause from consequence." In *The biology of rarity*, edited by W. E. Kunin and K. J. Gaston, 91–109. London: Chapman and Hall.

Chown, S. L., A. Addo-Bediako, and K. J. Gaston. 2002. "Physiological variation in insects: Large-scale patterns and their implications." *Comparative Biochemistry and Physiology B: Biochemistry and Molecular Biology* 131 (4): 587–602.

Chown, S. L., and K. J. Gaston. 1999. "Exploring links between physiology and ecology at macro-scales: The role of respiratory metabolism in insects." *Biological Reviews* 74 (1): 87–120.

———. 2010. "Body size variation in insects: A macroecological perspective." *Biological Reviews* 85:139–169.

Chown, S. L., and C. J. Klok. 2003. "Altitudinal body size clines: Latitudinal effects associated with changing seasonality." *Ecography* 26 (4): 445–455.

Chown, S. L. , E. Marais, M. D. Picker, and J. S. Terblanche. 2006. "Gas exchange characteristics, metabolic rate and water loss of the Heelwalker, *Karoophasma biedouwensis* (Mantophasmatodea: Austrophasmatidae)." *Journal of Insect Physiology* 52:442–449.

Chown, S. L., E. Marais, J. S. Terblanche, C. J. Klok, J. R. B. Lighton, and T. M. Blackburn. 2007. "Scaling of insect metabolic rate is inconsistent with the nutrient supply network model." *Functional Ecology* 21:282–290.

Chown, S. L., and S. W. Nicolson. 2004. *Insect physiological ecology: Mechanisms and patterns*. Oxford: Oxford University Press.

Chown, S. L., and H. E. Steenkamp. 1996. "Body size and abundance in a dung beetle assemblage: Optimal mass and the role of transients." *African Entomology* 4 (2): 203–212.

Chown, S. L., and J. S. Terblanche. 2007. "Physiological diversity in insects: Ecological and evolutionary contexts." *Advances in Insect Physiology* 33:50–152.

Cichoń, M., and J. Kozłowski. 2000. "Ageing and typical survivorship curves result from optimal resource allocation." *Evolutionary Ecology Research* 2 (7): 857–870.

Cohen, J. E., T. Jonsson, C. B. Müller, H. C. J. Godfray, and V. M. Savage. 2005. "Body sizes of hosts and parasitoids in individual feeding relationships." *Proceedings of the National Academy of Sciences of the United States of America* 102 (3): 684–689.

Colbo, M. H., and G. N. Porter. 1979. "The interaction of rearing temperature and food supply on the life history of two species of Simuliidae (Diptera)." *Canadian Journal of Zoology* 57:301–306.

Couvillon, M. J., J. M. Jandt, N. Duong, and A. Dornhaus. 2010. "Ontogeny of worker body size distribution in bumble bee (*Bombus impatiens*) colonies." *Ecological Entomology* 35:424–435.

Cushman, J. H., J. H. Lawton, and B. F. J. Manly. 1993. "Latitudinal patterns in European ant assemblages: Variation in species richness and body size." *Oecologia* 95 (1): 30–37.

Darveau, C. A. , Hochachka. P. W., Roubik. D. W., and R. K. Suarez. 2005a. "Allometric scaling of flight energetics in orchid bees: Evolution of flux capacities and flux rates." *Journal of Experimental Biology* 208:3593–3602.

Darveau, C. A., P. W. Hochachka, K. C. Welch, D. W. Roubik, and R. K. Suarez. 2005b. "Allometric scaling of flight energetics in Panamanian orchid bees: A comparative phylogenetic approach." *Journal of Experimental Biology* 208:3581–3591.

David, J. R., P. Gibert, E. Gravot, G. Petavy, J. P. Morin, D. Karan, and B. Moreteau. 1997. "Phenotypic plasticity and developmental temperature in Drosophila: Analysis and significance of reaction norms of morphometrical traits." *Journal of Thermal Biology* 22 (6): 441–451.

Davidowitz, G., D. A. Roff, and H. F. Nijhout. 2005. "A physiological perspective on the response of body size and development time to simultaneous directional selection." *Integrative and Comparative Biology* 45:525–531.

De Block, M., and R. Stoks. 2008. "Compensatory growth and oxidative stress in a damselfly." *Proceedings of the Royal Society B: Biological Sciences* 275:781–785.

de Jong, G., and T. M. van der Have. 2009. "Temperature dependence of development rate, growth rate and size: From biophysics to adaptation." In *Phenotypic plasticity of insects: Mechanisms and consequences*, edited by D. W. Whitman and T. N. Ananthakrishnan, 523–588. Enfield: Science Publishers.

den Boer, P. J. 1979. "Exclusion or coexistence and the taxonomic or ecological relationship between species." *Netherlands Journal of Zoology* 30:278–306.

DeVries, P. J., and R. Dudley. 1990. "Morphometrics, airspeed, thermoregulation, and lipid reserves of migrating *Urania fulgens* (Uraniidae) moths in natural free flight." *Physiological Zoology* 63 (1): 235–251.

Dial, K. P., and J. M. Marzluff. 1988. "Are the smallest organisms the most diverse?" *Ecology* 69:1620–1624.

Dillon, M. E., M. R. Frazier, and R. Dudley. 2006. "Into thin air: Physiology and evolution of alpine insects." *Integrative and Comparative Biology* 46:49–61.

Diniz-Filho, J. A. F., P. de Marco, and B. A. Hawkins. 2010. "Defying the curse of ignorance: Perspectives in insect macroecology and conservation." *Insect Conservation and Diversity* 3:172–179.

Diniz-Filho, J. A. F., and H. G. Fowler. 1998. "Honey ants (Genus*Myrmecocystus*) macroecology: Effect of spatial patterns on the relationship between worker body size and geographic range size." *Environmental Entomology* 27:1094–1101.

Dixon, A. F. G., and J. L. Hemptinne. 2001. "Body size distribution in predatory ladybird beetles reflects that of their prey." *Ecology* 82 (7): 1847–1856.

Dixon, A. F. G., P. Kindlmann, and V. Jarošik. 1995. "Body size distribution in aphids: Relative surface area of specific plant structures." *Ecological Entomology* 20 (2): 111–117.

Dolphin, K., and D. L. J. Quicke. 2001. "Estimating the global species richness of an incompletely described taxon: An example using parasitoid wasps (Hymenoptera: Braconidae)." *Biological Journal of the Linnean Society* 73 (3): 279–286. doi: 10.1006/Bij1.2001.0537.

Dudley, R. 1998. "Atmospheric oxygen, giant Paleozoic insects and the evolution of aerial locomotor performance." *Journal of Experimental Biology* 201 (8): 1043–1050.

———. 2000a. *The biomechanics of insect flight: Form, function, evolution.* Princeton: Princeton University Press.

———. 2000b. "The evolutionary physiology of animal flight: Paleobiological and present perspectives." *Annual Review of Physiology* 62:135–155.

Edgar, B. A. 2006. "How flies get their size: Genetics meets physiology." *Nature Reviews Genetics* 7:907–916.

Emlen, D. J. 1997a. "Alternative reproductive tactics and male-dimorphism in the horned beetle *Onthophagus acuminatus* (Coleoptera:Scarabaeidae)." *Behavioral Ecology and Sociobiology* 41 (5): 335–341.

———. 1997b. "Diet alters male horn allometry in the beetle *Onthophagus acuminatus* (Coleoptera: Scarabaeidae)." *Proceedings of the Royal Society B: Biological Sciences* 264 (1381): 567–574.

Emlen, D. J., and C. E. Allen. 2004. "Genotype to phenotype: Physiological control of trait size and scaling in insects." *Integrative and Comparative Biology* 43:617–634.

Emlen, D. J., J. Hunt, and L. W. Simmons. 2005. "Evolution of sexual dimorphism and male dimorphism in the expression of beetle horns: Phylogenetic evidence for modularity, evolutionary lability, and constraint." *American Naturalist* 166:S42–S68.

Emlen, D. J., L. C. Lavine, and B. Ewen-Campen. 2007. "On the origin and evolutionary diversification of beetle horns." *Proceedings of the National Academy of Sciences of the United States of America* 104 (suppl. 1): 8661–8668.

Emlen, D. J., and H. F. Nijhout. 2000. "The development and evolution of exaggerated morphologies in insects." *Annual Review of Entomology* 45: 661–708.

Ernsting, G., and J. A. Isaaks. 1997. "Effects of temperature and season on egg size, hatchling size and adult size in *Notiophilus biguttatus*." *Ecological Entomology* 22 (1): 32–40.

Espadaler, X., and C. Gómez. 2002. "The species body-size distribution in Iberian ants is parameter independent." *Vie et Milieu* 52 (2–3): 103–107.

Etienne, R. S., and H. Olff. 2004. "How dispersal limitation shapes species–body size distributions in local communities." *American Naturalist* 163 (1): 69–83.

Evans, E. W. 2000. "Morphology of invasion: Body size patterns associated with establishment of *Coccinella septempunctata* (Coleoptera: Coccinellidae) in western North America." *European Journal of Entomology* 97 (4): 469–474.

Fairbairn, D. J. 1992. "The origins of allometry: size and shape polymorphism in the common waterstrider, *Gerris remigis* Say (Heteroptera, Gerridae)." *Biological Journal of the Linnean Society* 45 (2): 167–186.

——. 1997. "Allometry for sexual size dimorphism: Pattern and process in the coevolution of body size in males and females." *Annual Review of Ecology and Systematics* 28:659–687.

——. 2005. "Allometry for sexual size dimorphism: Testing two hypotheses for Rensch's rule in the water strider*Aquarius remigis*." *American Naturalist* 166 (S69–S84).

Feener, D. H., Jr., J. R. B. Lighton, and G. A. Bartholomew. 1988. "Curvilinear allometry, energetics and foraging ecology: A comparison of leaf-cutting ants and army ants." *Functional Ecology* 2 (4): 509–520.

Fenchel, T., and B. J. Finlay. 2004. "The ubiquity of small species: Patterns of local and global diversity." *Bioscience* 54 (8): 777–784.

Finlay, B. J. 2002. "Global dispersal of free-living microbial eukaryote species." *Science* 296 (5570): 1061–1063.

Finlay, B. J., J. A. Thomas, G. C. McGavin, T. Fenchel, and R. T. Clarke. 2006. "Self-similar patterns of nature: Insect diversity at local to global scales." *Proceedings of the Royal Society B: Biological Sciences* 273:1935–1941.

Frankino, W. A., B. J. Zwaan, D. L. Stern, and P. M. Brakefield. 2005. "Natural selection and developmental constraints in the evolution of allometries." *Science* 307 (5710): 718–720.

Fraser, V. S. , B. Kaufmann, B. P. Oldroyd, and R. H. Crozier. 2000. "Genetic influence on caste in the ant*Camponotus consobrinus*." *Behavioural Ecology and Sociobiology* 47:188–194.

Frazier, M. R., H. A. Woods, and J. F. Harrison. 2001. "Interactive effects of rearing temperature and oxygen on the development of*Drosophila melanogaster*." *Physiological and Biochemical Zoology* 74 (5): 641–650.

Gahlhoff, J. E., Jr. 1998. *Smallest adult*. http://entnemdept.ufl.edu/walker/ufbir/files/pdf/UFBIR_Chapter38.pdf 1998. Available from http://entnemdept.ufl.edu/walker/ufbir/files/pdf/UFBIR_Chapter38.pdf.

García-Barros, E., and M. L. Munguira. 1997. "Uncertain branch lengths, taxonomic sampling error, and the egg to body size allometry in temperate butterflies (Lepidoptera)." *Biological Journal of the Linnean Society* 61 (2): 201–221.

Gardezi, T., and J. da Silva. 1999. "Diversity in relation to body size in mammals: A comparative study." *American Naturalist* 153 (1): 110–123.

Gardner, J. L., A. Peters, M. R. Kearney, L. Joseph, and R. Heinsohn. 2011. "Declining body size: A third universal response to warming?" *Trends in Ecology and Evolution* 26:285–291.

Gaston, K. J. 1988a. "The intrinsic rates of increase of insects of different sizes." *Ecological Entomology* 13 (4): 399–409.

———. 1988b. "Patterns in the local and regional dynamics of moth populations." *Oikos* 53 (1): 49–57.

———. 1991. "The magnitude of global insect species richness." *Conservation Biology* 5 (3): 283–296.

———. 1994. "Spatial patterns of species description: How is our knowledge of the global insect fauna growing." *Biological Conservation* 67 (1): 37–40.

Gaston, K. J., and T. M. Blackburn. 2000. *Pattern and process in macroecology.* Oxford: Blackwell Science.

Gaston, K. J., T. M. Blackburn, P. M. Hammond, and N. E. Stork. 1993. "Relationships between abundance and body size—where do tourists fit?" *Ecological Entomology* 18 (4): 310–314.

Gaston, K. J., S. L. Chown, and K. L. Evans. 2008. "Ecogeographic rules: Elements of a synthesis." *Journal of Biogeography* 35:483–500.

Gaston, K. J., S. L. Chown, and R. D. Mercer. 2001. "The animal species–body size distribution of Marion Island." *Proceedings of the National Academy of Sciences of the United States of America* 98 (25): 14493–14496.

Gaston, K. J., S. L. Chown, and C. V. Styles. 1997. "Changing size and changing enemies: The case of the mopane worm." *Acta Oecologica* 18 (1): 21–26.

Gaston, K. J., and E. Hudson. 1994. "Regional patterns of diversity and estimates of global insect species richness." *Biodiversity and Conservation* 3 (6): 493–500.

Gaston, K. J., and J. H. Lawton. 1988. "Patterns in the distribution and abundance of insect populations." *Nature* 331 (6158): 709–712.

Gaston, K. J., D. Reavey, and G. R. Valladares. 1991. "Changes in feeding habit as caterpillars grow." *Ecological Entomology* 16 (3): 339–344.

Gauld, I. D., and K. J. Gaston. 1995. "The Costa Rican Hymenoptera fauna." In *The Hymenoptera of Costa Rica*, edited by P. E. Hanson and I. D. Gauld, 13–19. Oxford: Oxford University Press.

Gaunt, M. W., and M. A. Miles. 2002. "An insect molecular clock dates the origin of the insects and accords with palaeontological and biogeographic landmarks." *Molecular Biology and Evolution* 19 (5): 748–761.

Gibbs, A. G., A. K. Chippindale, and M. R. Rose. 1997. "Physiological mechanisms of evolved desiccation resistance in *Drosophila melanogaster.*" *Journal of Experimental Biology* 200 (12): 1821–1832.

Gilchrist, G. W., and R. B. Huey. 2004. "Plastic and genetic variation in wing loading as a function of temperature within and among parallel clines in *Drosophila subobscura*." *Integrative and Comparative Biology* 44 (6): 461–470.

Glazier, D. S. 2005. "Beyond the '3/4-power law': Variation in the intra- and interspecific scaling of metabolic rate in animals." *Biological Reviews* 80:611–662.

——. 2010. "A unifying explanation for diverse metabolic scaling in animals and plants." *Biological Reviews* 85:111–118.

Gómez, C., and X. Espadaler. 2000. "Species body-size distribution and spatial scale in Iberian ants." *Vie et Milieu* 50 (4): 289–295.

Gotthard, K. 2004. "Growth strategies and optimal body size in temperate Pararginii butterflies." *Integrative and Comparative Biology* 44 (6): 471–479.

Gotthard, K., S. Nylin, and C. Wiklund. 2000. "Individual state controls temperature dependence in a butterfly (*Lasiommata maera*)." *Proceedings of the Royal Society B: Biological Sciences* 267 (1443): 589–593.

Gouws, E. J. 2007. "Intraspecific body size variation in insects." MSc thesis, Stellenbosch University, Stellenbosch.

Gouws, E. J., K. J. Gaston, and S. L. Chown. 2011. "Intraspecific body size frequency distributions of insects." *PLoS ONE* 6:e16606.

Gowing, G., and H. F. Recher. 1984. "Length-weight relationships for invertebrates from forests in south-eastern New South Wales." *Australian Journal of Ecology* 9 (1): 5–8.

Graham, J. B., R. Dudley, N. M. Aguilar, and C. Gans. 1995. "Implications of the late Paleozoic oxygen pulse for physiology and evolution." *Nature* 375 (6527): 117–120.

Grandchamp, A-C., J. Niemelä, and J. Kotze. 2000. "The effects of trampling on assemblages of ground beetles (Coleoptera, Carabidae) in urban forests in Helsinki, Finland." *Urban Ecosystems* 4 (4): 321–332.

Green, A. J. 1999. "Allometry of genitalia in insects and spiders: One size does not fit all." *Evolution* 53 (5): 1621–1624.

Greenlee, K. J., and J. F. Harrison. 2005. "Respiratory changes throughout ontogeny in the tobacco hornworm caterpillar, *Manduca sexta*." *Journal of Experimental Biology* 208:1385–1392.

Greve, M., K. J. Gaston, B. J. van Rensburg, and S. L. Chown. 2008. "Environmental factors, regional body size distributions and spatial variation in body size of local avian assemblages." *Global Ecology and Biogeography* 17:514–523.

Grimaldi, D., and M. S. Engel. 2005. *Evolution of the insects*. Cambridge: Cambridge University Press.

Gutiérrez, D., and R. Menéndez. 1997. "Patterns in the distribution, abundance and body size of carabid beetles (Coleoptera: Caraboidea) in relation to dispersal ability." *Journal of Biogeography* 24 (6): 903–914.

Hanski, I., and Y. Cambefort. 1991. "Resource partitioning." In *Dung beetle ecology*, edited by I. Hanski and Y. Cambefort, 330–349. Princeton: Princeton University Press.

Haro, R. J., K. Edley, and M. J. Wiley. 1994. "Body-size and sex-ratio in emergent stonefly nymphs (*Isogenoides olivaceus*, Perlodidae): Variation between cohorts and populations." *Canadian Journal of Zoology* 72 (8): 1371–1375.

Harrison, J. F., A. Kaiser, and VandenBrooks J. M. 2010. "Atmospheric oxygen level and the evolution of insect body size." *Proceedings of the Royal Society B: Biological Sciences* 277:1937–1946.

Hawkins, B. A. 1995. "Latitudinal body-size gradients for the bees of the eastern United States." *Ecological Entomology* 20 (2): 195–198.

Hawkins, B. A., and J. H. Lawton. 1995. "Latitudinal gradients in butterfly body sizes: Is there a general pattern?" *Oecologia* 102 (1): 31–36.

Heinrich, B., and G. A. Bartholomew. 1979. "Roles of endothermy and size in inter- and intraspecific competition for elephant dung in an African dung beetle, *Scarabaeus laevistriatus*." *Physiological Zoology* 52 (4): 484–496.

Hemmingsen, A. M. 1934. "A statistical analysis of the differences in body size of related species." *Videnskabelige Meddelelser fra Dansk Naturhistorik Forening i Kobenhavn* 98:125–160.

Hodkinson, I. D., and D. S. Casson. 2000. "Patterns within patterns: Abundance-size relationships within the Hemiptera of tropical rain forest or why phylogeny matters." *Oikos* 88 (3): 509–514.

Honěk, A. 1993. "Intraspecific variation in body size and fecundity in insects: A general relationship." *Oikos* 66 (3): 483–492.

——. 1999. "Constraints on thermal requirements for insect development." *Entomological Science* 2:615–621.

Hu, Y., Y. Xie, F. Zhu, C. Wang, and C. Lei. 2010. "Variation in sexual size dimorphism among populations: Testing the differential-plasticity hypothesis." *Entomologia Experimentalis et Applicata* (137): 204–209.

——. 2011. "Development time and body size in Eupolyphaga sinensis along a latitudinal gradient from China." *Environmental Entomology* 40:1–7.

Huey, R. B., G. W. Gilchrist, M. L. Carlson, D. Berrigan, and L. Serra. 2000. "Rapid evolution of a geographic cline in size in an introduced fly." *Science* 287 (5451): 308–309.

Huey, R. B., and P. D. Ward. 2005. "Hypoxia, global warming, and terrestrial late Permian extinctions." *Science* 308:398–401.

Hughes, W. O. H., S. Sumner, S. Van Borm, and J. J. Boomsma. 2003. "Worker caste polymorphism has a genetic basis in *Acromyrmex* leaf-cutting ants." *Proceedings of the National Academy of Sciences of the United States of America* 100:9394–9397.

Huston, M. A., and S. Wolverton. 2011. "Regulation of animal size by eNPP, Bergmann's rule, and related phenomena." *Ecological Monographs* 81:349–405.

Hutchinson, G. E., and R. H. MacArthur. 1959. "A theoretical ecological model of size distributions among species of animals." *American Naturalist* 93:117–125.

Irlich, U. M., J. S. Terblanche, T. M. Blackburn, and S. L. Chown. 2009. "Insect rate-temperature relationships: Environmental variation and the metabolic theory of ecology." *American Naturalist* 174:819–835.

James, A. C., R. B. R. Azevedo, and L. Partridge. 1995. "Cellular basis and developmental timing in a size cline of *Drosophila melanogaster*." *Genetics* 140 (2): 659–666.

Janzen, D. H. 1973. "Sweep samples of tropical foliage insects: Description of study sites, with data on species abundances and size distributions." *Ecology* 54 (3): 659–686.

Janzen, D. H., M. Ataroff, M. Fariñas, S. Reyes, N. Rincon, A. Soler, P. Soriano, and M. Vera. 1976. "Changes in the arthropod community along an elevational transect in the Venezuelan Andes." *Biotropica* 8 (3): 193–203.

Johnson, K. P., S. F. Bush, and D. H. Clayton. 2005. "Correlated evolution of host and parasite body size: Tests of Harrison's rule using birds and lice." *Evolution* 29:1744–1753.

Johst, K., and R. Brandl. 1997. "Body size and extinction risk in a stochastic environment." *Oikos* 78 (3): 612–617.

Juliano, S. A. 1985. "The effects of body size on mating and reproduction in *Brachinus lateralis* (Coleoptera: Carabidae)." *Ecological Entomology* 10 (3): 271–280.

Kaiser, A., C. J. Klok, J. J. Socha, W. K. Lee, M. C. Quinlan, and J. F. Harrison. 2007. "Increase in tracheal investment with beetle size supports hypothesis of oxygen limitation on insect gigantism." *Proceedings of the National Academy of Sciences of the United States of America* 104:13198–13203.

Kari, J. S., and R. B. Huey. 2000. "Size and seasonal temperature in free-ranging *Drosophila subobscura*." *Journal of Thermal Biology* 25 (4): 267–272.

Kaspari, M. 1993. "Body sze and microclimate use in Neotropical granivorous ants." *Oecologia* 96 (4): 500–507.

Kaspari, M., and E. L. Vargo. 1995. "Colony size as a buffer against seasonality: Bergmann's rule in social insects." *American Naturalist* 145 (4): 610–632.

Kaspari, M., and M. D. Weiser. 1999. "The size-grain hypothesis and interspecific scaling in ants." *Functional Ecology* 13 (4): 530–538.

Katzourakis, A., A. Purvis, S. Azmeh, G. Rotheray, and F. Gilbert. 2001. "Macroevolution of hoverflies (Diptera : Syrphidae): The effect of using higher-level taxa in studies of biodiversity, and correlates of species richness." *Journal of Evolutionary Biology* 14 (2): 219–227.

Kause, A., I. Saloniemi, J. P. Morin, E. Haukioja, S. Hanhimaki, and K. Ruohomaki. 2001. "Seasonally varying diet quality and the quantitative genetics of development time and body size in birch feeding insects." *Evolution* 55 (10): 1992–2001.

Kennington, W. J., and A. A. Hoffmann. 2010. "The genetic architecture of wing size divergence at varying spatial scales along a body size cline in *Drosophila melanogaster*." *Evolution* 64:1935–1943.

Kennington, W. J., A. A. Hoffmann, and L. Partridge. 2007. "Mapping regions within cosmopolitan inversion In(3R)Payne associated with natural variation in body size in *Drosophila melanogaster*." *Genetics* 177 (1): 549–556.

Kindlmann, P., A. F. G. Dixon, and I. Dostalkova. 1999. "Does body size optimization result in skewed body size distribution on a logarithmic scale?" *American Naturalist* 153 (4): 445–447.

Kingsolver, J. G., and D. W. Pfennig. 2004. "Individual-level selection as a cause of Cope's rule of phyletic size increase." *Evolution* 58 (7): 1608–1612.

Kirk, W. D. J. 1991. "The size relationship between insects and their hosts." *Ecological Entomology* 16 (3): 351–359.

Klass, K.-D., M. D. Picker, J. Damgaard, S. van Noort, and K. Tojo. 2003. "The taxonomy, genitalic morphology, and phylogenetic relationships of Southern African Mantophasmatodea (Insecta)." *Entomologische Abhandlungen* 61 (1): 3–67.

Klok, C. J., and S. L. Chown. 1999. "Assessing the benefits of aggregation: thermal biology and water relations of anomalous emperor moth caterpillars." *Functional Ecology* 13 (3): 417–427.

Klok, C. J., and J. F. Harrison. 2009. "Atmospheric hypoxia limits selection for large body size in insects." *PLoS ONE* 4:e3876.

Klok, C. J., A. J. Hubb, and J. F. Harrison. 2009. "Single and multigenerational responses of body mass to atmospheric oxygen concentrations in *Drosophila melanogaster*: Evidence for roles of plasticity and evolution." *Journal of Evolutionary Biology* 22:2496–2504.

Klok, C. J., A. Kaiser, J. R. B. Lighton, and J. F. Harrison. 2010. "Critical oxygen partial pressures and maximal tracheal conductances for *Drosophila melanogaster* reared for multiple generations in hypoxia or hyperoxia." *Journal of Insect Physiology* 56:461–469.

Koleff, P., K. J. Gaston, and J. J. Lennon. 2003. "Measuring beta diversity for presence-absence data." *Journal of Animal Ecology* 72 (3): 367–382.

Kooijman, S. A. L. M. 2010. *Dynamic energy budget theory for metabolic organization*. Cambridge: Cambridge University Press.

Kottelat, M., R. Britz, T. H. Hui, and K. -E. Witte. 2006. "*Paedocypris*, a new genus of Southeast Asian cyprinid fish with a remarkable sexual dimorphism, comprises the world's smallest vertebrate." *Proceedings of the Royal Society B: Biological Sciences* 273:895–899.

Kozłowski, J. 1992. "Optimal allocation of resources to growth and reproduction: Implications for age and size at maturity." *Trends in Ecology and Evolution* 7 (1): 15–19.

Kozłowski, J., M. Czarnołęski, and M. Dańko. 2004. "Can optimal resource allocation models explain why ectotherms grow larger in cold?" *Integrative and Comparative Biology* 44 (6): 480–493.

Kozłowski, J., and A. T. Gawelczyk. 2002. "Why are species' body size distributions usually skewed to the right?" *Functional Ecology* 16 (4): 419–432.

Kozłowski, J., M. Konarzewski, and A. T. Gawelczyk. 2003. "Cell size as a link between noncoding DNA and metabolic rate scaling." *Proceedings of the National Academy of Sciences of the United States of America* 100 (24): 14080–14085.

Kozłowski, J., and A. T. Teriokhin. 1999. "Allocation of energy between growth and reproduction: The Pontryagin maximum principle solution for the case of age- and season-dependent mortality." *Evolutionary Ecology Research* 1 (4): 423–441.

Kozłowski, J., and J. Weiner. 1997. "Interspecific allometries are by-products of body size optimization." *American Naturalist* 149 (2): 352–380.

Kozłowski, J., and R. G. Wiegert. 1987. "Optimal age and size at maturity in annuals and perennials with determinate growth." *Evolutionary Ecology* 1:231–244.

Krüger, O., and G. C. McGavin. 2000. "Macroecology of local insect communities." *Acta Oecologica—International Journal of Ecology* 21 (1): 21–28.

Kukalová-Peck, J. 1985. "Ephemeroid wing venation based upon new gigantic carboniferous mayflies and basic morphology, phylogeny, and metamorphosis of pterygote insects (Insecta, Ephemerida)." *Canadian Journal of Zoology* 63 (4): 933–955.

———. 1987. "New carboniferous Diplura, Monura, and Thysanura, the hexapod ground plan, and the role of thoracic side lobes in the origin of wings (Insecta)." *Canadian Journal of Zoology* 65 (10): 2327–2345.

Labandeira, C. C., and J. J. Sepkoski. 1993. "Insect diversity in the fossil record." *Science* 261 (5119): 310–315.

Lawton, J. H., and K. C. Brown. 1986. "The population and community ecology of invading insects." *Philosophical Transactions of the Royal Society B: Biological Sciences* 314 (1167): 607–617.

Lease, H. M., and B. O. Wolf. 2010. "Exoskeletal chitin scales isometrically with body size in terrestrial insects." *Journal of Morphology* 271:759–768.

———. 2011. "Lipid content of terrestrial arthropods in relation to body size, phylogeny, ontogeny and sex." *Physiological Entomology* 36:29–38.

Lee, S. F., L. Rako, and A. A. Hoffmann. 2011. "Genetic mapping of adaptive wing size variation in *Drosophila simulans*." *Heredity* 107:22–29.

Liebherr, J. K. 1988. "Brachyptery and phyletic size increase in Carabidae (Coleoptera)." *Annals of the Entomological Society of America* 81 (2): 157–163.

Lighton, J. R. B., M. C. Quinlan, and D. H. Feener, Jr. 1994. "Is bigger better? Water balance in the polymorphic desert harvester ant *Messor pergandei*." *Physiological Entomology* 19 (4): 325–334.

Loder, N. 1997. "Insect species–body size distributions." PhD thesis, University of Sheffield.

Loder, N., T. M. Blackburn, and K. J. Gaston. 1997. "The slippery slope: Towards an understanding of the body size frequency distribution." *Oikos* 78 (1): 195–201.

Loeschcke, V., J. Bundgaard, and J. S. F. Barker. 2000. "Variation in body size and life history traits in *Drosophila aldrichi* and *D. buzzatii* from a latitudinal cline in eastern Australia." *Heredity* 85 (5): 423–433.

Lounibos, L. P. 1994. "Geographical and developmental components of adult size of Neotropical Anopheles (Nyssorhynchus)." *Ecological Entomology* 19 (2): 138–146.

Makarieva, A. M., V. G. Gorshkov, and B. L. Li. 2005. "Temperature-associated upper limits to body size in terrestrial poikilotherms." *Oikos* 111:425–436.

Mangel, M., and S. B. Munch. 2005. "A life-history perspective on short- and long-term consequences of compensatory growth." *American Naturalist* 166:E155–E176.

Marden, J. H. 1995. "Large-scale changes in thermal sensitivity of flight performance during adult maturation in a dragonfly." *Journal of Experimental Biology* 198 (10): 2095–2102.

Margraf, N., K. Gotthard, and M. Rahier. 2003. "The growth strategy of an alpine beetle: Maximization or individual growth adjustment in relation to seasonal time horizons?" *Functional Ecology* 17 (5): 605–610.

Masaki, S. 1996. "Geographical variation of life cycle in crickets (Ensifera: Grylloidea)." *European Journal of Entomology* 93 (3): 281–302.

Maurer, B. A. 1998. "The evolution of body size in birds. I. Evidence for non-random diversification." *Evolutionary Ecology* 12 (8): 925–934.

———. 2003. "Adaptive diversification of body size: The roles of physical constraint, energetics and natural selection." In *Macroecology: Concepts and consequences*, edited by T. M. Blackburn and K. J. Gaston, 174. Oxford: Blackwell Science.

Maurer, B. A., J. H. Brown, and R. D. Rusler. 1992. "The micro and macro in body size evolution." *Evolution* 46 (4): 939–953.

May, R. M. 1978. "The dynamics and diversity of insect faunas." In *Diversity of insect faunas*, edited by L. A. Mound and N. Waloff. New York: Blackwell Scientific Publications.

———. 2000. "The dimensions of life on earth." In *Nature and human society: The quest for a sustainable world*, edited by P. H. Raven and T. Williams, 30. Washington, DC: National Academy Press.

McKinney, M. L. 1990. "Trends in body size evolution." In *Evolutionary trends*, edited by K. J. McNamara, 75–118. Tucson: University of Arizona Press.

Mercer, R. D., A. G. A. Gabriel, J. Barendse, D. J. Marshall, and S. L. Chown. 2001. "Invertebrate body sizes from Marion Island." *Antarctic Science* 13 (2): 135–143.

Miller, P. L. 1966. "Supply of oxygen to active flight muscles of some large beetles." *Journal of Experimental Biology* 45 (2): 285–304.

Mirth, C. K., and L. M. Riddiford. 2007. "Size assessment and growth control: How adult size is determined in insects." *BioEssays* 29:344–355.

Moczek, A. P. 2006. "A matter of measurements: Challenges and approaches in the comparative analysis of static allometries." *American Naturalist* 167:606–611.

Moczek, A. P., and H. F. Nijhout. 2003. "Rapid evolution of a polyphenic threshold." *Evolution and Development* 5:259–268.

Moreteau, B., P. Gibert, J. M. Delpuech, G. Petavy, and J. R. David. 2003. "Phenotypic plasticity of sternopleural bristle number in temperate and tropical populations of *Drosophila melanogaster.*" *Genetical Research* 81 (1): 25–32.

Morse, D. R., J. H. Lawton, M. M. Dodson, and M. H. Williamson. 1985. "Fractal dimension of vegetation and the distribution of arthropod body lengths." *Nature* 314 (6013): 731–733.

Morse, D. R., N. E. Stork, and J. H. Lawton. 1988. "Species number, species abundance and body length relationships of arboreal beetles in Bornean lowland rain-forest trees." *Ecological Entomology* 13 (1): 25–37.

Myburgh, E. 2001. "The influence of developmental temperature on the survival of adult *Simulium chutteri* (Diptera: Simuliidae)." MSc thesis, University of Pretoria.

Nijhout, H. F. 2003. "The control of body size in insects." *Developmental Biology* 261:1–9.

Nijhout, H. F., G. Davidowitz, and D. A. Roff. 2006. "A quantitative analysis of the mechanism that controls body size in *Manduca sexta.*" *Journal of Biology* 5:1–15.

Nijhout, H. F., D. A. Roff, and G. Davidowitz. 2010. "Conflicting processes in the evolution of body size and development time." *Philosophical Transactions of the Royal Society B: Biological Sciences* 365:567–575.

Novotný, V., and Y. Basset. 1999. "Body size and host plant specialization: A relationship from a community of herbivorous insects on *Ficus* from Papua New Guinea." *Journal of Tropical Ecology* 15:315–328.

Novotný, V., Y. Basset, S. E. Miller, G. D. Weiblen, B. Bremer, L. Cizek, and P. Drozd. 2002. "Low host specificity of herbivorous insects in a tropical forest." *Nature* 416 (6883): 841–844.

Novotný, V., and P. Kindlmann. 1996. "Distribution of body sizes in arthropod taxa and communities." *Oikos* 75 (1): 75–82.

Novotný, V., and M. R. Wilson. 1997. "Why are there no small species among xylem-sucking insects?" *Evolutionary Ecology* 11 (4): 419–437.

Nygren, G. H., A. Bergström, and S. Nylin. 2008. "Latitudinal body size clines in the butterfly *Polyommatus icarus* are shaped by gene-environment interactions." *Journal of Insect Science* 8:Art. 47.

Nylin, S., and K. Gotthard. 1998. "Plasticity in life-history traits." *Annual Review of Entomology* 43:63–83.

Nylin, S., K. Gotthard, and C. Wiklund. 1996. "Reaction norms for age and size at maturity in *Lasiommata* butterflies: Predictions and tests." *Evolution* 50:1351–1358.

Ødegaard, F., O. H. Diserud, S. Engen, and K. Aagaard. 2000. "The magnitude of local host specificity for phytophagous insects and its implications for estimates of global species richness." *Conservation Biology* 14 (4): 1182–1186.

Olalla-Tárraga, M. Á. 2011. "'Nullius in Bergmann' or the pluralistic approach to ecogeographical rules: A reply to Watt et al. (2010)." *Oikos* 120:1441–1444.

Olalla-Tárraga, M. Á., and M. Á. Rodríguez. 2007. "Energy and interspecific body size patterns of amphibian faunas in Europe and North America: Anurans follow Bergmann's rule, urodeles its converse." *Global Ecology and Biogeography* 16:606–617.

Orme, C. D. L., N. J. B. Isaac, and A. Purvis. 2002a. "Are most species small? Not within species-level phylogenies." *Proceedings of the Royal Society B: Biological Sciences* 269 (1497): 1279–1287.

Orme, C. D. L., D. L. J. Quicke, J. M. Cook, and A. Purvis. 2002b. "Body size does not predict species richness among the metazoan phyla." *Journal of Evolutionary Biology* 15 (2): 235–247.

Oster, G. F., and E. O. Wilson. 1978. *Caste and ecology in the social insects.* Princeton: Princeton University Press.

Partridge, L., B. Barrie, K. Fowler, and V. French. 1994. "Evolution and development of body-size and cell-size in *Drosophila melanogaster* in response to temperature." *Evolution* 48 (4): 1269–1276.

Partridge, L., and J. A. Coyne. 1997. "Bergmann's rule in ectotherms: Is it adaptive?" *Evolution* 51 (2): 632–635.

Peat, J., B. Darvill, J. Ellis, and D. Goulson. 2005. "Effects of climate on intra- and interspecific size variation in bumblebees." *Functional Ecology* 19 (1): 145–151.

Peck, L. S., and S. H. P. Maddrell. 2005. "Limitation of size by hypoxia in the fruit fly *Drosophila melanogaster*." *Journal of Experimental Zoology* 303A:968–975.

Picker, M. D., B. Leon, and J. G. H. Londt. 1991. "The hypertrophied hindwings of *Palmipenna aeoleoptera* Picker, 1987 (Neuroptera, Nemopteridae) reduce attack by robber flies by increasing apparent body size." *Animal Behaviour* 42:821–825.

Pielou, E. C. 1975. *Ecological diversity.* Wiley-Interscience: New York.

Porter, E. E., and B. A. Hawkins. 2001. "Latitudinal gradients in colony size for social insects: Termites and ants show different patterns." *American Naturalist* 157 (1): 97–106.

Pörtner, H. O. 2001. "Climate change and temperature-dependent biogeography: Oxygen limitation of thermal tolerance in animals." *Naturwissenschaften* 88:137–146.

Powell, S., and N. R. Franks. 2006. "Ecology and the evolution of worker morphological diversity: A comparative analysis with *Eciton* army ants." *Functional Ecology* 20:1105–1114.

Prange, H. D., and B. Pinshow. 1994. "Thermoregulation of an unusual grasshopper in a desert environment: Ihe importance of food source and body size." *Journal of Thermal Biology* 19 (1): 75–78.

Rascón, B., and J. F. Harrison. 2005. "Oxygen partial pressure effects on metabolic rate and behavior of tethered flying locusts." *Journal of Insect Physiology* 51:1193–1199.

Reichle, D. 1968. "Relation of body size to food intake $O_2$ consumption and trace element metabolism in forest floor arthropods." *Ecology* 49 (3): 538–542.

Ribera, I., S. Dolédec, I. S. Downie, and G. N. Foster. 2001. "Effect of land disturbance and stress on species traits of ground beetle assemblages." *Ecology* 82 (4): 1112–1129.

Roberts, S. P., J. F. Harrison, and R. Dudley. 2004. "Allometry of kinematics and energetics in carpenter bees (*Xylocopa varipuncta*) hovering in variable-density gases." *Journal of Experimental Biology* 207 (6): 993–1004.

Roces, F., and J. R. B. Lighton. 1995. "Larger bites of leaf-cutting ants." *Nature* 373 (6513): 392–393.

Rockstein, M., and J. Miquel. 1973. "Aging in insects." In *The physiology of insecta*, vol. 1, 2nd ed., edited by M. Rockstein, 371–478. New York: Academic Press.

Röder, G., M. Rahier, and R. E. Naisbit. 2008. "Counter-intuitive developmental plasticity induced by host quality." *Proceedings of the Royal Society B: Biological Sciences* 275:879–885.

Rodrigues, D., and G. R. P. Moreira. 2004. "Seasonal variation in larval host plants and consequences for *Heliconius erato* (Lepidoptera : Nymphalidae) adult body size." *Austral Ecology* 29 (4): 437–445.

Roff, D. 1980. "Optimizing development time in a seasonal environment: The ups and downs of clinal variation." *Oecologia* 45 (2): 202–208.

Roff, D. A. 1992. *The evolution of life histories: Theory and analysis.* London: Chapman and Hall.

———. 2001. "Age and size at maturity." In *Evolutionary ecology: Concepts and case studies*, edited by C. W. Fox, D. A. Roff, and D. J. Fairbairn, 99–112. Oxford: Oxford University Press.

———. 2002. *Life history evolution.* Sunderland, MA: Sinauer Associates

Rogers, D. J., and S. E. Randolph. 1991. "Mortality rates and population-density of tsetse flies correlated with satellite imagery." *Nature* 351 (6329): 739–741.

Rogers, L. E., W. T. Hinds, and R. L. Buschbom. 1976. "General weight vs. length relationship for insects." *Annals of the Entomological Society of America* 69 (2): 387–389.

Roulston, T. H., and J. H. Cane. 2000. "The effect of diet breadth and nesting ecology on body size variation in bees (Apiformes)." *Journal of the Kansas Entomological Society* 73:129–142.

Rowland, J. M., and D. J. Emlen. 2009. "Two thresholds, three male forms result in facultative male trimorphism in beetles." *Science* 323:773–776.

Sample, B. E., R. J. Cooper, R. D. Greer, and R. C. Whitmore. 1993. "Estimation of insect biomass by length and width." *American Midland Naturalist* 129 (2): 234–240.

Schöning, C., W. Kinuthia, and N. R. Franks. 2005. "Evolution of allometries in the worker caste of *Dorylus* army ants." *Oikos* 110:231–240.

Scriber, J. M. 2002. "Latitudinal and local geographic mosaics in host plant preferences as shaped by thermal units and voltinism in *Papilio* spp. (Lepidoptera)." *European Journal of Entomology* 99 (2): 225–239.

Sequeira, R., and M. Mackauer. 1993. "Seasonal variation in body size and offspring sex ratio in field populations of the parasitoid wasp, *Aphidius ervi* (Hymenoptera: Aphidiidae)." *Oikos* 68:340–346.

Shear, W. A., and J. Kukalová-Peck. 1990. "The ecology of Paleozoic terrestrial arthropods: The fossil evidence." *Canadian Journal of Zoology* 68 (9): 1807–1834.

Shingleton, A. W., C. K. Mirth, and P. W. Bates. 2008. "Developmental model of static allometry in holometabolous insects." *Proceedings of the Royal Society B: Biological Sciences* 275:1875–1885.

Siemann, E., J. Haarstad, and D. Tilman. 1999a. "Dynamics of plant and arthropod diversity during old field succession." *Ecography* 22 (4): 406–414.

Siemann, E., D. Tilman, and J. Haarstad. 1999b. "Abundance, diversity and body size: Patterns from a grassland arthropod community." *Journal of Animal Ecology* 68 (4): 824–835.

Simmons, L. W., and T. J. Ridsdill-Smith. 2011. *Ecology and evolution of dung beetles*. Chichester: Wiley-Blackwell.

Smith, F. A., J. H. Brown, J. P. Haskell, S. K. Lyons, J. Alroy, E. L. Charnov, T. Dayan, B. J. Enquist, S. K. M. Ernest, E. A. Hadly, K. E. Jones, D. M. Kaufman, P. A. Marquet, B. A. Maurer, K. J. Niklas, W. P. Porter, B. Tiffney, and M. R. Willig. 2004. "Similarity of mammalian body size across the taxonomic hierarchy and across space and time." *American Naturalist* 163 (5): 672–691.

Smith, R. J., A. Hines, S. Richmond, M. Merrick, A. Drew, and R. Fargo. 2000. "Altitudinal variation in body size and population density of *Nicrophorus investigator* (Coleoptera : Silphidae)." *Environmental Entomology* 29 (2): 290–298.

Snelling, E. P., R. S. Seymour, P. G. D. Matthews, S. Runciman, and C. R. White. 2011. "Scaling of resting and maximum hopping metabolic rate throughout the life cycle of *Locusta migratoria*." *Journal of Experimental Biology* 214:3218–3224.

Spicer, J. I., and K. J. Gaston. 1999. *Physiological diversity and its ecological implications*. Oxford: Blackwell Science.

Stearns, S. C. 1992. *The evolution of life histories*. Oxford: Oxford University Press.

Steffan-Dewenter, I., and T. Tscharntke. 1997. "Early succession of butterfly and plant communities on set-aside fields." *Oecologia* 109 (2): 294–302.

Stevens, M. M., S. Jackson, S. A. Bester, J. S. Terblanche, and S. L. Chown. 2010. "Oxygen limitation and thermal tolerance in two terrestrial arthropod species." *Journal of Experimental Biology* 213:2209–2218.

Stevenson, R. D. 1985. "The relative importance of behavioural and physiological adjustments controlling body temperature in terrestrial ectotherms." *American Naturalist* 126 (3): 362–386.

Stillwell, R. C., W. U. Blanckenhorn, T. Teder, G. Davidowitz, and C. W. Fox. 2010. "Sex differences in phenotypic plasticity affect variation in sexual size dimorphism in insects: From physiology to evolution." *Annual Review of Entomology* 55:227–245.

Stillwell, R. C., and G. Davidowitz. 2010a. "A developmental perspective on the evolution of sexual size dimorphism of a moth." *Proceedings of the Royal Society B: Biological Sciences* 277:2069–2074.

———. 2010b. "Sex difference in phenotypic plasticity of a mechanism that controls body size: Implications for sexual size dimorphism." *Proceedings of the Royal Society B: Biological Sciences* 277:3819–3826.

Stillwell, R. C., and C. W. Fox. 2009. "Geographic variation in body size, sexual size dimorphism and fitness components of a seed beetle: Local adaptation versus phenotypic plasticity." *Oikos* 118:703–712.

Stillwell, R. C., G. E. Morse, and C. W. Fox. 2007. "Geographic variation in body size and sexual size dimorphism of a seed-feeding beetle." *American Naturalist* 170:358–369.

Stoks, R., M. De Block, and M. A. McPeek. 2006a. "Physiological costs of compensatory growth in a damselfly." *Ecology* 87:1566–1574.

Stoks, R., M. De Block, W. W. Van Dooerslaer, and J. Rolff. 2006b. "Time constraints mediate predator-induced plasticity in immune function, condition, and life history." *Ecology* 87:809–815.

Stone, G. N. 1994. "Activity patterns of females of the solitary bee *Anthophora plumipes* in relation to temperature, nectar supplies and body size." *Ecological Entomology* 19 (2): 177–189.

Stone, G. N., P. M. J. Loder, and T. M. Blackburn. 1995. "Foraging and courtship behaviour in males of the solitary bee *Anthophora plumipes* (Hymenoptera:

Anthophoridae): Thermal physiology and the roles of body size." *Ecological Entomology* 20 (2): 169–183.
Stone, G. N., and P. G. Willmer. 1989. "Warm-up rates and body temperatures in bees: The importance of body size, thermal regime and phylogeny." *Journal of Experimental Biology* 147:303–328.
Strobbe, F., and R. Stoks. 2004. "Life history reaction norms to time constraints in a damselfly: Differential effects on size and mass." *Biological Journal of the Linnean Society* 83 (2): 187–196.
Tammaru, T., P. Kaitaniemi, and K. Ruohomäki. 1996. "Realized fecundity in *Epirrita autumnata* (Lepidoptera: Geometridae): Relation to body size and consequences to population dynamics." *Oikos* 77 (3): 407–416.
Tantawy, A. O. 1964. "Studies on natural populations of *Drosophila*. III. Morphological and genetic differences of wing length in *Drosophila melanogaster* and *D. simulans* in relation to season." *Evolution* 18 (4): 560–570.
Taylor, Brad W., Chester R. Anderson, and Barbara L. Peckarsky. 1998. "Effects of size at metamorphosis on stonefly fecundity, longevity, and reproductive success." *Oecologia* 114 (4): 494–502.
Teder, T., and T. Tammaru. 2005. "Sexual size dimorphism within species increases with body size in insects." *Oikos* 108 (2): 321–334.
Teder, T., T. Tammaru, and T. Esperk. 2008. "Dependence of phenotypic variance in body size on environmental quality." *American Naturalist* 172:223–232.
Tomkins, J. L., and G. S. Brown. 2004. "Population density drives the local evolution of a threshold dimorphism." *Nature* 431 (7012): 1099–1103.
Tomkins, J. L., and W. Hazel. 2011. "Explaining phenotypic diversity: The conditional strategy and threshold trait expression." In *Ecology and evolution of dung beetles*, edited by L. W. Simmons and T. J. Ridsdill-Smith. Chichester: Wiley-Blackwell.
Tomkins, J. L., J. S. Kotiaho, and N. R. Lebas. 2005. "Matters of scale: Positive allometry and the evolution of male dimorphisms." *American Naturalist* 165:389–402.
———. 2006. "Major differences in minor allometries: A reply to Moczek." *American Naturalist* 167:612–618.
Tompkins, D. M., and D. H. Clayton. 1999. "Host resources govern the specificity of swiftlet lice: Size matters." *Journal of Animal Ecology* 68:489–500.
Tseng, M. 2003. "Life-history responses of a mayfly to seasonal constraints and predation risk." *Ecological Entomology* 28 (1): 119–123.
Ulrich, W. 2004. "Allometric ecological distributions in a local community of Hymenoptera." *Acta Oecologica* 25 (3): 179–186.
van der Have, T. M., and G. de Jong. 1996. "Adult size in ectotherms: Temperature effects on growth and differentiation." *Journal of Theoretical Biology* 183 (3): 329–340.
Verberk, W. C. E. P., D. T. Bilton, P. Calosi, and J. I. Spicer. 2011. "Oxygen sup-

ply in aquatic ectotherms: Partial pressure and solubility together explain biodiversity and size patterns." *Ecology* 92:1565–1572.

Waddington, K. D., L. H. Herbst, and D. W. Roubik. 1986. "Relationship between recruitment systems of stingless bees and within-nest worker size variation." *Journal of the Kansas Entomological Society* 59:95–102.

Walters, R. J., and M. Hassall. 2006. "The temperature-size rule in ectotherms: May a general explanation exist after all? ." *American Naturalist* 167: 510–523.

Warren, M., M. A. McGeoch, S. W. Nicolson, and S. L. Chown. 2006. "Body size patterns in *Drosophila* inhabiting a mesocosm: Interactive effects of spatial variation in temperature and abundance." *Oecologia* 149:245–255.

Wasserman, S. S., and C. Mitter. 1978. "The relationship of body size to breadth of diet in some Lepidoptera." *Ecological Entomology* 3 (2): 155–160.

Watt, C., S. Mitchell, and V. Salewski. 2010. "Bergmann's rule: A concept cluster?" *Oikos* 119:89–100.

Watt, C., and V. Salewski. 2011. "Bergmann's rule encompasses mechanism: A reply to Olalla-Tárraga (2001)." *Oikos* 120:1445–1447.

Weeks, A. R., S. W. McKechnie, and A. A. Hoffmann. 2002. "Dissecting adaptive clinal variation: Markers, inversions and size/stress associations in *Drosophila melanogaster* from a central field population." *Ecology Letters* 5 (6): 756–763.

West, G. B., J. H. Brown, and B. J. Enquist. 1997. "A general model for the origin of allometric scaling laws in biology." *Science* 276 (5309): 122–126.

White, C. R., T. M. Blackburn, J. S. Terblanche, E. Marais, M. Gibernau, and S. L. Chown. 2007. "Evolutionary responses of discontinuous gas exchange in insects." *Proceedings of the National Academy of Sciences of the United States of America* 104:8357–8361.

Whitman, D. W. 2008. "The significance of body size in Orthoptera: A review." *Journal of Orthoptera Research* 17:117–134.

Whitman, D. W., and T. N. Ananthakrishnan. 2009. *Phenotypic plasticity of insects: Mechanisms and consequences.* Enfield: Science Publishers.

Wilf, P., C. C. Labandeira, K. R. Johnson, and B. Ellis. 2006. "Decoupled plant and insect diversity after the end-cretaceous extinction." *Science* 313: 1112–1115.

Williams, D. M. 2001. *Largest.* http://entnemdept.ufl.edu/walker/ufbir/chapters/chapter_30.shtml 2001. Available from http://entnemdept.ufl.edu/walker/ufbir/chapters/chapter_30.shtml.

Williamson, M., and K. J. Gaston. 2005. "The lognormal distribution is not an appropriate null hypothesis for the species-abundance distribution." *Journal of Animal Ecology* 74 (3): 409–422.

Willmer, P. G., and D. M. Unwin. 1981. "Field analyses of insect heat budgets: Reflectance, size and heating rates." *Oecologia* 50 (2): 250–255.

Woods, H. A., W. Makino, J. B. Cotner, S. E. Hobbie, J. F. Harrison, K. Acharya, and J. J. Elser. 2003. "Temperature and the chemical composition of poikilothermic organisms." *Functional Ecology* 17 (2): 237–245.

Wootton, R. J., and J. Kukalová-Peck. 2000. "Flight adaptations in Palaeozoic Palaeoptera (Insecta)." *Biological Reviews of the Cambridge Philosophical Society* 75 (1): 129–167.

Yang, A. S., C. H. Martin, and H. F. Nijhout. 2004. "Geographic variation of caste structure among ant populations." *Current Biology* 14:514–519.

Yom-Tov, Y., and E. Geffen. 2011. "Recent spatial and temporal changes in body size of terrestrial vertebrates: Probable causes and pitfalls." *Biological Reviews* 86:531–541.

Yuval, B. , J. A. Wekesa, D. Lemenager, E. E. Kauffman, and R. K. Washino. 1993. "Seasonal variation in body size of mosquitoes (Diptera: Culicidae) in a rice culture agroecosystem." *Environmental Entomology* 22:459–463.

CHAPTER TWO

# Latitudinal Variation of Body Size in Land Snail Populations and Communities

Jeffrey C. Nekola, Gary M. Barker, Robert A. D. Cameron, and Beata M. Pokryszko

Bergmann's rule states the tendency for body size to positively correlate with latitude within groups of closely related animals (Bergmann 1847; Rensch 1938; Mayr 1956; James 1970). While there has been much discussion about the appropriate taxonomic groups and scales where this pattern is expected to operate, there is little question that it applies to most endothermic vertebrates (e.g., Blackburn et al. 1999). However, not only are the underlying mechanisms still the subject of considerable debate (Ashton et al. 2000), but it is also less clear if this pattern should be expected for ectothermic invertebrates. This is because while larger ectothermic bodies experience lowered heat loss rates due to smaller surface-to-volume ratios, they will also experience correspondingly lower heat absorption rates (Cushman et al. 1993).

It should thus not be surprising that general body size vs. latitude relationships have not been forthcoming for these species. While positive body size–latitude relationships have been identified in ants (Cushman et al. 1993; Kaspari and Vargo 1995), marine isopods (Poulin 1995b), copepods (Poulin 1995a), amphipods (Poulin and Hamilton 1995), and monogoneans (Poulin 1996), marine bivalves tend to have largest body sizes at midlatitudes (Roy and Martien 2001). While Northern Hemisphere butterflies possess positive body size correlations with increasing latitude, Southern Hemisphere taxa demonstrate a negative relationship (Barlow 1994; Hawkins and Lawton 1995). Eastern North American bees demonstrate no latitudinal patterns in body size within state faunas and five of eight investigated families. Of the three families that do demonstrate significant patterns, two actually become smaller with increasing latitude (Hawkins 1995). Regional land snail faunas in northwest Europe also demonstrate a negative correlation between body size and latitude, though this pattern may be related to phylogenetic constraints (Hausdorf 2003).

A potential limitation of these prior analyses is that local faunas (i.e., 100 × 100 km regions or greater) are typically generated from geographic range maps. This leads to two important potential sources of error. First, all taxa will contribute equally to calculated statistics, no matter their actual abundance in that local region. This will bias results in favor of rare species, which will typically be more frequent than common taxa in a metacommunity (Hubbell 2001). Second, as species will never completely saturate their range (Hurlbert and White 2005), it should not be assumed that all mapped taxa will actually be present within a given region or locality. One way to address these problems is to investigate body size relationships at the grain of individual populations or communities, where empirical determination of body sizes and composition patterns is possible via field observations.

This chapter documents the nature and underlying mechanisms for body size vs. latitude relationships in land snail populations and communities observed across large extents (>2,500 km) in northwestern Europe, eastern North America, and New Zealand. As these regions possess little phylogenetic overlap (even at higher taxonomic levels), they represent fairly independent tests for latitudinal control of body size. Interregional comparisons are appropriate, however, as each supports similarly rich community assemblages (e.g., >30 taxa/0.1 ha; Barker 2005; Nekola 2005; Pokryszko and Cameron 2005). The following four questions will be addressed: (1) Does individual body size within a species positively correlate with latitude? (2) Are communities of high latitude represented by a greater proportion of large taxa? (3) Are communities of high latitude represented by a greater proportion of large individuals? (4) Does altitudinal variation in community body size mimic latitudinal relationships?

## Materials and Methods

### *Datasets*

Within-species analysis of body size vs. latitude was conducted on seven North American taxa: *Carychium exiguum, C. exile, Gastrocopta procera, G. rogersensis, Cochlicopa lubrica, C. lubricella*, and *C. morseana*, representing 99–5,177 individuals from 25–126 sites spread across 472–1,094 km extents (table 2.1). These taxa represent three families (Carychiidae, Pupillidae, and Cochlicopidae) within two orders (Basommatophora and Stylommatophora; Hubricht 1985). Specific methods used to sample these populations are provided in Nekola and Coles (2001) and Nekola and Barthel (2002). The height and width of all undamaged adult shells collected from each population was measured in increments of 0.02 mm using a dissecting microscope with a calibrated ocular micrometer. The latitude-longitude coordinates for each population were determined through digitization of USGS (or equivalent) 7.5 minute topographic maps, or use of a handheld GPS unit. To minimize statistical bias from use of polar coordinates, site locations were converted to Cartesian UTM Zone 16 coordinates using ARCINFO.

Analysis of community body size vs. latitude patterns was based on 128–2,476 sites sampled within each landscape (table 2.1). European samples extended from Scotland and western Ireland to Ukraine and Finland. In this region data were available at two scales. Species presence/absence data from the 46 areas discussed in Pokryszko and Cameron (2005) were used to analyze size class frequency variation within site faunas. Species abundance data from the 128 sites used to generate most of the above species lists were used to document size class frequency variation within the individuals encountered at each site. North American samples extended from the western shore of Hudson's Bay to the Ozark Plateau, New England, and the South Carolina coastal plain. New Zealand samples were collected throughout the archipelago, including the outlying island systems of the Antipodes, Auckland, Bounty, Campbell, Chatham, Kermadec, Snares, and Three Kings. The latitude-longitude coordinates for each site were determined through digitization of topographic maps or use of a handheld GPS. As before, data were converted to Cartesian UTM coordinates using ARCINFO. To allow more direct comparison of Northern and Southern Hemisphere latitudinal patterns, UTM northing coordinates for the New Zealand sites were

TABLE 2.1 **Overview of Datasets Used in Analyses**

a. Within Species

| Species | No. of Sites | No. of Individuals | Latitudinal Extent (km) | Source |
|---|---|---|---|---|
| *Carychium exiguum* | 35 | 1,178 | 586 | Nekola and Barthel 2002 |
| *Carychium exile* | 116 | 5,177 | 1,094 | Nekola and Barthel 2002 |
| *Gastrocopta procera* | 24 | 343 | 997 | Nekola and Coles 2001 |
| *Gastrocopta rogersensis* | 25 | 415 | 955 | Nekola and Coles 2001 |
| *Cochlicopa lubrica* | 108 | 961 | 764 | Nekola unpublished |
| *Cochlicopa lubricella* | 126 | 621 | 780 | Nekola unpublished |
| *Cochlicopa moreseana* | 25 | 99 | 472 | Nekola unpublished |

b. Within Communities—Latitude

| Region | No. of Sites | No. of Taxa | No. of Individuals | Latitudinal Extent | Total Extent | Source |
|---|---|---|---|---|---|---|
| Europe | 128 | 140 | 70,175 | 1,676 | 2,778 | Pokryszko and Cameron 2005 |
| North America | 898 | 167 | 423,322 | 1,771 | 2,508 | Nekola 2005 |
| New Zealand | 2,476 | 790 | 2,379,991 | 2,557 | 2,807 | Baker 2005 |

c. Within Communities—Altitude

| Region | No. of Sites | No. of Taxa | No. of Individuals | Altitudinal Extent | Source |
|---|---|---|---|---|---|
| New Mexico, USA | 120 | 45 | 23,434 | 2,050 | Dillon and Metcalf 1997 |

converted to UTM southing coordinates by subtraction of each from 9,997,823 m.

Specific methods used to document communities at each site are presented in Pokryszko and Cameron (2005), Nekola (2005), and Barker (2005). Although minor methodological variations exist, in each region all communities were analyzed through standardized litter sampling augmented by hand searching for larger taxa within approximately 0.1 ha areas, as outlined in Cameron and Pokryszko (2005). All recovered identifiable shells were assigned to species, subspecies, or morphospecies.

Approximate adult shell height and width for each encountered taxon were determined from median sizes reported by Kerney et al. (1983), Pilsbry (1948), Nekola (unpublished data) and Barker (unpublished data). Each taxon was then placed into one of five size classes based on maximum shell dimension: micro (≤ 2 mm), minute (2.1–5 mm), small (5.1–10 mm), medium (10.1–20 mm), and large (>20 mm). To better reflect the faunas of these landscapes, size classes were modified from Emberton (1995) by breaking his "minute" (<5 mm) class in two, and by lumping together his "large" (20.1–40 mm) and "giant" (>40.1 mm) classes.

The contribution of each size class to the taxa and individuals present within each community was then determined. The contribution of a given size class to the taxa observed on a given site was calculated by dividing the number of taxa falling into each size class by the site's total richness. Contribution of a given size class to the individuals encountered on a given site was calculated by dividing the abundance of individuals falling within each size class by the total number of encountered individuals from that site. These metrics may provide slightly different insights. While the taxa-based metric will be a more conservative measure because of the greater replicatability of taxa lists vs. abundance data within sites (Cameron and Pokryszko 2005), the individual-based metric may provide better insight into the nature of environmental control, as each individual represents a separate sample unit.

Because of the similarity often observed between latitudinal and altitudinal body size relationships (Hausdorf 2003), analyses were also conducted on the variation in size class frequency within site taxa lists and individual abundances along elevation gradients in New Mexico (USA) mountain land snail communities using data presented in Dillon and Metcalf (1997). Communities were sampled at 61 m elevation intervals from 1,700–3,600 m in the Mogollon, Sierra Blanca, Sangre de Cristo, and Mt. Taylor ranges, and at 90 m intervals from 1,600–2,900 m along the Rio

Peñasco and Tularosa River in the Sacramento Mountains. Data analysis was conducted as for latitudinal relationships, with maximum shell dimension being based on median size data reported in Pilsbry (1948).

### *Statistical Procedures*

As all analyzed relationships appeared approximately linear, with variance along the best-fit lines being both normal and homoscedastic, least-squares linear regression was used to characterize variation with latitude/altitude in: (1) shell size within species; and (2) size class frequencies for taxa/individuals within sites. As five regressions were performed separately on each community dataset (one for each size class), a Bonferroni correction was used to adjust the critical value in these analyses to $p = 0.01$. Because of differences in substrate, landscape position, and sampling intervals in the seven New Mexican elevational transects, a mixed model was used to block transect site effect prior to analysis of altitudinal patterns. In this analysis, partial $r^2$ values of size class frequency vs. altitude, given transect effects, were reported.

Because the European datasets exist at two sampling scales (46 presence/absence lists generated from summation of multiple sites collected within a locality, and abundance lists from 128 separate sites), these data may not be easily comparable to the North American and New Zealand datasets, in which taxa lists and abundances are known from each site. To attempt to compensate for this discrepancy, sites from North America and New Zealand were geographically grouped into 300 × 300 km square regions. Average size class frequencies were then calculated within each region, and regression statistics were repeated on these average responses.

## Results

### *Latitudinal Size Variation within Species*

Only one of the seven analyzed taxa (*Gastrocopta rogersensis*) did not demonstrate significant ($p = 0.876$) variation in size with latitude (fig.2.1; table 2.2). Of the remaining taxa, four (*Carychium exiguum, C. exile, Cochlicopa lubrica, C. lubricella*) exhibited significant ($p < 0.0005$) increases in shell size with latitude, while the remaining two (*Gastrocopta procera, Cochlicopa morseana*) demonstrated significant ($p < 0.0005$)

decreases. The amount of variation accounted for by latitude varied from 14% (*Carychium exile*) to 38% (*Gastrocopta procera*), while the magnitude of changes ranged from approximately 10% of shell height per 1,000 km in *Cochlicopa morseana* to 22% in *Carychium exile*.

Latitudinal variation of five North American land snails

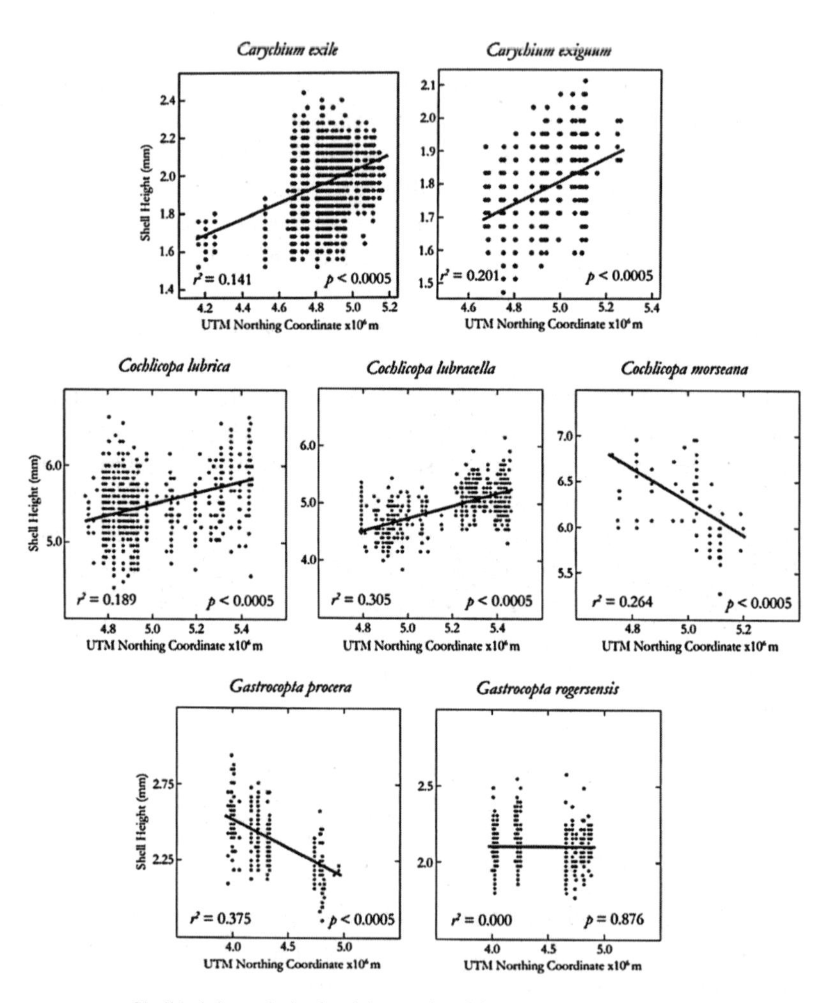


FIGURE 2.1. Shell height vs. latitude with associated least-squares regression statistics and best-fit lines for six North American land snail species: *Carychium exile, Carychium exiguum, Cochlicopa lubrica, Cochlicopa lubracella, Cochlicopa morseana, Gastrocopta procera*, and *Gastrocopta rogersensis*.

TABLE 2.2 **Latitudinal Variation in Adult Shell Height for Seven North American Land Snail Species**

| Species | Average Height (mm) | Change per 1,000 km N | Percent Change per 1,000 km N | $r^2$ | $p$ |
|---|---|---|---|---|---|
| *Carychium exiguum* | 1.95 | 0.347 | 19.6 | 0.201 | <0.0005 |
| *Carychium exile* | 1.77 | 0.424 | 21.7 | 0.141 | <0.0005 |
| *Gastrocopta procera* | 2.40 | −0.367 | −15.3 | 0.375 | <0.0005 |
| *Gastrocopta rogersensis* | 2.11 | −0.003 | −0.1 | 0.000 | 0.876 |
| *Cochlicopa lubrica* | 5.50 | 0.738 | 13.4 | 0.189 | <0.0005 |
| *Cochlicopa lubracella* | 4.95 | 0.964 | 19.4 | 0.305 | <0.0005 |
| *Cochlicopa morseana* | 6.29 | −0.609 | −9.7 | 0.264 | <0.0005 |

### *Latitudinal Variation of Size Class Frequencies within Communities*

In northwestern Europe the frequency of micro and minute taxa demonstrated significant ($p < 0.0005$ and $p = 0.002$) positive correlations with latitude, while the frequency of medium and large taxa showed significant ($p < 0.0005$ and $p = 0.001$) negative correlations (table 2.3). Latitude explained 20%–30% of observed variation in these size classes. The frequency of small-sized taxa did not demonstrate a statistically significant ($p = 0.098$) correlation with latitude. While southern sites were dominated by medium taxa (ca. 40% of a site fauna list), northern sites were characterized by minute (ca. 30%), micro (ca. 30%), and small (20%) taxa (fig. 2.2). The frequency of minute northwestern European individuals also demonstrated a significant ($p < 0.0005$) positive correlation with latitude while the frequency of medium and large individuals both exhibited significant ($p < 0.0005$) negative correlations (table 2.4). Latitude explained 10%–40% of observed variation in these size classes. The frequency of both micro and small-sized individuals demonstrated no significant ($p = 0.534$ and $p = 0.276$) correlation with latitude. While micro and medium individuals dominated southern sites (each representing ca. 35% of total), northern sites were characterized by minute (ca. 45%) and micro (35%) individuals (fig. 2.2).

In eastern North America the frequency of minute taxa demonstrated a significant ($p < 0.0005$) positive correlation with latitude, while the frequency of micro, medium, and large taxa showed significant ($p < 0.0005$) negative correlations (table 2.3). Latitude explained 1.6%–3.6% of observed variation in these size classes, with up to 24% of observed variation being accounted for by latitude when sites were grouped into 300 × 300 km regions. The frequency of small-sized taxa did not demonstrate a

TABLE 2.3 **Variation in Shell Size Class Frequency of Taxa within Site Faunas of Increasing Latitude**

| Region and Size Class | Constant | Slope | $r^2$ | $p$ |
|---|---|---|---|---|
| Northwestern Europe: | | | | |
| Micro | −20.805 | 5.961 | 0.302 | <0.0005 |
| Minute | −6.966 | 5.513 | 0.203 | 0.002 |
| Small | 6.287 | 3.418 | 0.061 | 0.098 |
| Medium | 101.923 | −12.083 | 0.254 | <0.0005 |
| Large | 19.561 | −2.808 | 0.232 | 0.001 |
| Eastern North America: | | | | |
| Micro | 48.211 | −4.198 | 0.016 | <0.0005 |
| Minute | 15.647 | 6.547 | 0.034 | <0.0005 |
| Small | 11.670 | 1.390 | 0.003 | 0.120 |
| Medium | 11.014 | −1.824 | 0.036 | <0.0005 |
| Large | 13.458 | −1.916 | 0.020 | <0.0005 |
| New Zealand: | | | | |
| Micro | −10.700 | 8.920 | 0.060 | <0.0005 |
| Minute | 60.449 | −0.821 | 0.001 | 0.255 |
| Small | 43.356 | −7.735 | 0.074 | <0.0005 |
| Medium | −0.448 | 1.195 | 0.008 | <0.0005 |
| Large | 7.344 | −1.559 | 0.031 | <0.0005 |

*Note*: Values are summary statistics generated via least-squares linear regression, with values scaled by $10^3$ km.

significant ($p = 0.12$) correlation with latitude. Southern sites were most often represented by minute (ca. 40%) and micro (ca. 30%) taxa, while northern sites were dominated by minute (ca. 60%) taxa (fig. 2.2). The frequency of minute eastern North American individuals also exhibited a significant ($p < 0.0005$) positive correlation with latitude, while the frequency of micro and medium individuals demonstrated significant ($p < 0.0005$ and $p = 0.001$) negative correlations (table 2.4). Latitude explained 1.4%–5.4% of observed variation in these size classes, with up to 24% of variation being accounted for by latitude when sites were grouped into 300 × 300 km regions. The frequency of both small and large-sized individuals demonstrated marginal correlations with latitude ($p = 0.026$ and $p = 0.012$), with small individuals tending to increase and large individuals tending to decrease in frequency with increasing latitude. Micro and minute individuals dominated the communities of southern sites (ca. 55% and 30% of total), while northern sites were characterized by minute (ca. 60%) and small (20%) individuals (fig. 2.2).

In New Zealand the frequency of micro and medium taxa exhibited significant ($p < 0.0005$ and $p = 0.008$) positive correlations with latitude, while the frequency of small and large taxa demonstrated significant ($p < 0.0005$) negative correlations (table 2.3). Latitude explained 1%–

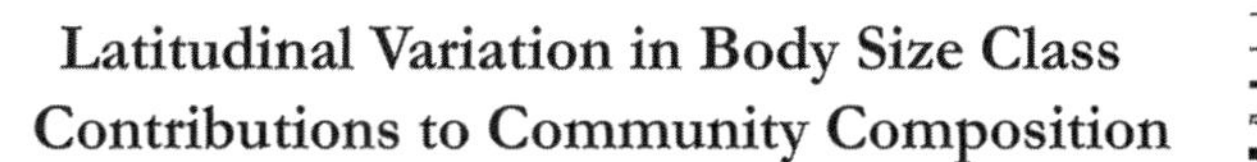


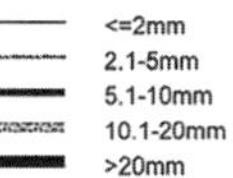


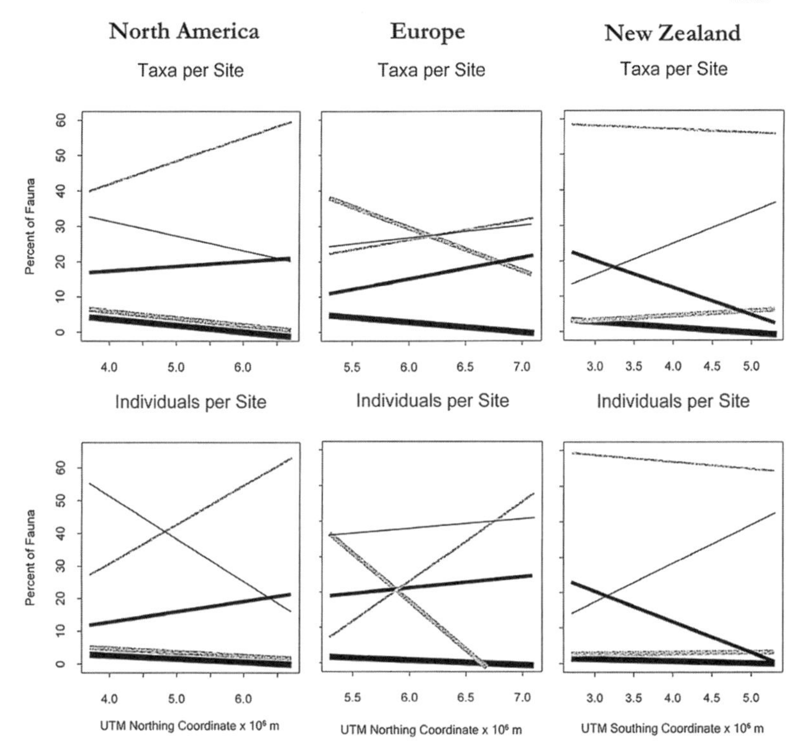


FIGURE 2.2. Average change in the contribution of five shell size classes (<2 mm = micro; 2.1–5 mm = minute; 5.1–10 mm = small; 10.1–20 mm = medium; >20 mm = large) for northwestern European, eastern North American, and New Zealand land snail community composition with increasing latitude, as represented by best-fit lines generated from least-squares linear regression. *a*, Variation in the frequency of size classes among taxa. *b*, Variation in the frequency of size classes among individuals.

7.4% of observed variation in these size classes, with up to 14% of variation being accounted for by latitude when sites were grouped into 300 × 300 km regions. The frequency of minute taxa did not demonstrate significant ($p = 0.255$) correlations with latitude. While northern sites were most often represented by minute (ca. 60%) and small (ca. 25%) taxa, southern sites were dominated by minute (ca. 60%) and micro (ca. 35%) taxa (fig. 2.2). The frequency of New Zealand micro individuals demonstrated a significant ($p < 0.0005$) positive correlation

TABLE 2.4 **Variation in Shell Size Class Frequency of Individuals within Site Faunas of Increasing Latitude**

| Region and Size Class | Constant | Slope | $r^2$ | $p$ |
|---|---|---|---|---|
| Northwestern Europe: | | | | |
| Micro | 21.155 | 2.799 | 0.003 | 0.534 |
| Minute | −112.570 | 22.601 | 0.363 | <0.0005 |
| Small | 2.040 | 3.162 | 0.009 | 0.276 |
| Medium | 180.745 | −27.238 | 0.386 | <0.0005 |
| Large | 8.629 | −1.323 | 0.105 | <0.0005 |
| Eastern North America: | | | | |
| Micro | 103.593 | −13.067 | 0.054 | <0.0005 |
| Minute | −16.882 | 11.912 | 0.054 | <0.0005 |
| Small | 0.017 | 3.164 | 0.006 | 0.026 |
| Medium | 5.892 | −0.979 | 0.014 | 0.001 |
| Large | 7.371 | −1.029 | 0.007 | 0.012 |
| New Zealand: | | | | |
| Micro | −15.174 | 10.853 | 0.047 | <0.0005 |
| Minute | 63.989 | −1.823 | 0.001 | 0.058 |
| Small | 45.992 | −8.584 | 0.062 | <0.0005 |
| Medium | 2.309 | 0.135 | 0.000 | 0.599 |
| Large | 2.882 | −0.580 | 0.011 | <0.0005 |

*Note*: Values are summary statistics generated via least-squares linear regression, with values scaled by 103 km.

with latitude while the frequency of small and large individuals exhibited significant ($p < 0.0005$) negative correlations. Latitude explained 1%–6.2% of observed variation in these size classes, with up to 10% of variation being accounted for by latitude when sites were grouped into 300 × 300 km regions. The frequency of both minute and medium individuals did not significantly ($p = 0.058$ and $p = 0.599$) correlate with latitude. While minute and small individuals dominated northern sites (ca. 60% and 25% of total), southern sites were characterized by minute (ca. 55%) and micro (40%) individuals (fig. 2.2).

### *Altitudinal Variation of Size Class Frequencies within Communities*

Few significant elevational trends in size class frequency of either taxa or individuals were noted within New Mexican mountain faunas (table 2.5). Only the micro class demonstrated significant variation, with these taxa becoming marginally less frequent ($p = 0.013$), and individuals becoming less frequent ($p = 0.003$) with increasing altitude. This altitudinal relationship explained an additional 7.1% and 10.1% of observed variation, following blocking for transect effects.

TABLE 2.5 **Variation in Shell Frequency with Increasing Elevation (Scaled by $10^3$ m) in Seven New Mexico, USA, Mountain Transects**

| Size Class | Constant | Slope | Partial $r^2$ | $p$ |
|---|---|---|---|---|
| a. Taxa within faunas: | | | | |
| Micro | 22.518 | −5.280 | 0.071 | 0.013 |
| Minute | 35.447 | 4.308 | 0.001 | 0.782 |
| Small | 18.327 | 3.929 | 0.019 | 0.208 |
| Medium | 26.229 | −2.013 | 0.005 | 0.532 |
| Large | 1.120 | 0.621 | 0.037 | 0.075 |
| b. Individuals within faunas: | | | | |
| Micro | 27.073 | −9.418 | 0.101 | 0.003 |
| Minute | 37.600 | 5.607 | 0.009 | 0.384 |
| Small | 14.633 | 6.613 | 0.021 | 0.178 |
| Medium | 21.084 | −2.974 | 0.008 | 0.421 |
| Large | −0.393 | 0.174 | 0.021 | 0.184 |

*Note*: Values are summary statistics based on mixed GLM models where site effects were blocked. Partial $r^2$ values represent the additional explanatory contribution of elevation following blocking of site effects.

## Discussion

In contrast to the positive relationship between body size and latitude posited by Bergmann's rule, land snails exhibited strong but inconsistent patterns. Positive, negative, and no correlations with latitude were observed within species. At the community level, analyses from the three disparate regions demonstrated that small taxa and individuals tend to increase in frequency, whereas large taxa and individuals tend to decrease in frequency with increasing latitude.

### *Body Size Patterns within Species*

The lack of consistent body size relationships with latitude mirrors previous studies that have been unable to identify general environmental correlates to land snail shell size. Shell size variation in *Albinaria idaea* and *Albinaria terebra* from Crete appear uncorrelated with contemporaneous precipitation, temperature/insolation, substrate chemistry, and elevation (Welter-Schultes 2000, 2001). However, a historical component appears important, because populations with greatest size variation are found in the oldest habitats (Welter-Schultes 2001). Lack of correlation between shell size and current environmental conditions, but apparent correlation of variance with population age, has also been noted for *Helix aspersa* in North Africa (Madec et al. 2003). Goodfriend (1986) could not identify universal ecological predictors for shell size in an extensive literature review, with individualistic responses being noted along moisture, temperature/insolation, and calcium availability gradients. He did document, however, positive correlations between shell size and moisture in ten of twelve taxa. Similar moisture vs. body size patterns have been more recently noted for *Cochlicopa lubrica* in Europe (Armbruster 2001) and in the fauna of the Kikai Islands of Japan over a 40 ka time frame (Hayakaze and Chiba 1999; Marui et al. 2004).

Cushman et al. (1993) suggest that positive body size correlations with latitude in ectotherms may be related to a greater capacity for larger organisms to store food, allowing for increased starvation resistance. This may be important in seasonal climates where individuals are forced to experience prolonged periods of low resources. The generality of this mechanism for land snails, however, is unclear as water may be a more important limiting resource than food (Riddle 1983; Goodfriend 1986). This would suggest that body size should also increase with frequency and severity of moisture stress. While this hypothesis cannot be tested via the current data, it is interesting to note that this prediction runs counter to the moisture vs. body size trends discussed above. Additionally, large snails in the arid southwestern USA are typically restricted to montane "islands" of higher and more regular precipitation, while lowland taxa that experience increased water stress tend to be of minute or micro size (e.g., Bequaert and Miller 1973; Metcalf and Smartt 1997). Starvation avoidance may thus not be an important general factor influencing land snail body size.

Turner and Lennon (1989) suggest that because of limited resources, smaller ectotherms will be more likely to maintain minimally viable populations at high latitudes, thus leading to a negative correlation between body size and latitude. This mechanism also does not apply to these data, as neither *Cochlicopa morseana* nor *Gastrocopta procrea* demonstrated a positive correlation between population size and latitude. Additionally, both of these taxa have closely related siblings that demonstrated positive body size correlations with latitude, even though these siblings presumably share similar resource demands and minimum viable population sizes. What does differentiate these two taxa from their siblings, however, is their lower latitude of geographic range (Hubricht 1985). As will be discussed below, if these species have a more limited ability to depress their freezing point, increased supercooling and winter survival abilities will be imparted by a reduction in body size.

The only consistent predictor of shell size identified by Goodfriend

(1986) was population density, which has been found to correlate inversely with adult shell size in European *Arianta arbustorum, Candidula intersecta, Cepaea nemoralis, Cochlicella acuta*, and *Helicella itala*. This relationship has also been observed in North American *Mesodon normalis* (Foster and Stiven 1996; Stiven and Foster 1996), though no impact was found on fecundity rates (Foster and Stiven 1994). Increased environmental heavy metal concentration may also correlate negatively with shell size in some species (Tryjanowski and Koralewska-Batura 2000).

These factors, too, are probably not responsible for the body size variation observed in the present study. All but one of the investigated species demonstrated population sizes that were independent of latitude (Nekola, unpublished data). And the lone taxon that did show a significant trend (*Cochlicopa lubricella*) had both significantly larger populations and shell sizes at higher latitudes. Additionally, heavy metal concentrations do not monotonically vary across eastern North America: pollution levels peak at midlatitudes along the industrial corridor extending from Chicago to New York.

### *Body Size Patterns between Communities*

Contrary to Bergmann's rule, across all three regions small-sized taxa and individuals tended to become more frequent at higher latitude, while larger-sized taxa and individuals tended to demonstrate a negative correlation. Variation in size class frequency of individuals within communities tended to follow the patterns observed for taxa. This general pattern was expressed somewhat differently in each region. The magnitude of micro and/or minute taxa increase and medium and/or large taxa decrease with increasing latitude was strongest in northwestern Europe. This general pattern differed in eastern North America, with the frequency of micro species decreasing at higher latitudes, and in New Zealand, with the frequency of medium species increasing at higher latitudes.

These variations do not invalidate the general pattern, as they probably represent differences in sampling protocols, regional ecological histories, gradients, and/or component phylogenies. The greater magnitude of variation in frequency of medium-sized taxa/individuals in Europe appears at least partially related to phylogenetic constraint due to the presence of the Clausiliidae, which has undergone significant radiation in the east and south of Europe (Kerney et al. 1983; Pokryszko and Cameron 2005). This family and its morphological niche (medium-sized individuals that are taller than wide) are essentially absent from both eastern North America and New Zealand. The stronger correlations with latitude found in Europe (as compared to the other two regions) are probably due to a more narrowed sampling regime, with only upland woods being observed. The lack of change in frequency for minute taxa/individuals in New Zealand appears related to in situ radiation of the Charopidae and Punctidae, which generated large numbers of minute taxa throughout the archipelago (Barker 2005). Additionally, the positive correlation between frequency of medium-sized taxa and latitude in New Zealand at best only weakly influences community composition, with less than 1% of observed variation being accounted for by this relationship. Additionally, medium-sized individuals did not demonstrate a corresponding significant correlation with latitude. The reduction of frequency of micro taxa/individuals in eastern North American communities, however, is not a phylogenetic or statistical artifact, as this size class includes taxa from seven different families, and as the relationship is equally significant for taxa and individuals. A possible explanation for this unique pattern is that only the North American dataset includes continental arctic climates. Thus, the highest-latitude northwestern Europe and New Zealand sites will be exposed to comparatively less severe winter temperatures.

Generally no body size vs. elevation relationships were noted from New Mexican montane faunas. Micro-sized taxa and individuals tended to decrease in frequency with increasing elevation, mirroring the latitudinal pattern seen in the eastern North American fauna. The lack of strong correlation between altitude and snail body size is not surprising, however, because in New Mexico elevation is strongly negatively correlated with precipitation levels (Metcalf and Smartt 1997). The confounding of elevation and precipitation in this region makes it unlikely that simple body size correlations will exist.

Two non-mutually-exclusive mechanisms can be advanced to account for the positive correlation between frequency of smaller-sized taxa/individuals in land snail communities and latitude: increased winter survival and dispersal ability with decreasing shell size.

Land snails have, at best, only a moderate ability to lower their freezing temperature. Many species apparently do not have cryoprotective chemicals (Riddle 1983), while those that do can only depress freezing temperatures to a limited degree (Ansart et al. 2002; Ansart 2002). For example, even though larger *Helix aspersa* individuals are able to depress freezing temperature more than small individuals, all perished within 16 hours of exposure to –50 C temperatures (Ansart and Vernon 2004). The principal avenue used by land snails to survive freezing is an increase in supercooling ability via gut clearance, prevention of debris contact with mantle tissue, and reduction of an individual's water mass (Riddle 1983). This last factor is probably the most important in terms of body size relationships, as the greater the water volume, the greater the likelihood that ice nucleation will occur, leading to freezing of tissues and mortality (Lee and Costanzo 1998; Ansart and Vernon 2003). Smaller land snails with lower water mass thus have greater abilities to supercool than larger individuals (Riddle and Miller 1988; Ansart and Vernon 2003). This pattern is common to most ectothermic invertebrates (Lee and Costanzo 1998). Additionally, because small snails experience greater water loss through their shells than larger taxa (Arad and Avivi 1998; Arad et al. 1998), they will have a greater ability to decrease water mass prior to the winter season.

Differential migration abilities of small vs. large snails, in combination with Pleistocene glaciation history, may also underlie this pattern. Active dispersal rates for land snail individuals are very low ($10^1$–$10^2$ m/yr; Hausdorf and Hennig 2003) with individuals being unable to cross barriers of $10^2$–$10^3$ m (Baur 1988; Schilthuizen and Lombaerts 1994). While there is a positive correlation between body size and distance moved over a single day, this relationship disappears by day 2 (Popov and Kramarenko 2004). Landscape migration of land snails is thus solely related to passive dispersal abilities (Hausdorf and Hennig 2003). Smaller-sized individuals have greater capabilities of long-range passive dispersal by wind (Kirchner et al. 1997) and animal (Pfenninger 2004) vectors. Additionally, while small snails have a lower fecundity per individual, they have much higher reproductive outputs when calculated on a per-volume basis. For example, while *Arianta arbustorum* individuals (16 × 21 mm) will lay 800 eggs during a lifetime and *Punctum pygmaeum* (0.7 × 1.4 mm) only six, *Punctum* produces 50 times the number of eggs per standard cubic unit of animal mass (data from Heller 2001). Such increased fecundity will increase the number of individuals capable of passive dispersal. Also, as the frequency of uniparental reproduction is inversely correlated with body size (Pokryszko and Cameron 2005), smaller taxa have the capability of founding new colonies via dispersal of only single individuals into new stations. Small-sized snails would thus be expected to be the most rapid colonizers of previously glaciated terrain, accounting for the positive correlation between the frequency of small snails in communities and latitude. A similar mechanism has been invoked to explain the disproportionate presence of small-shelled taxa as colonists of Pacific islands (Vagvolgyi 1975; Marui et al. 2004). Large-size snails of Pacific islands have largely developed from in situ evolution of small taxa into unfilled niches (Cowie 1995).

## Conclusions

These analyses demonstrate that (1) general body size relationships with latitude do not exist within a species; (2) communities of high latitude tend to be represented by a greater proportion of small taxa; (3) communities of high latitude tend to be represented by a greater proportion of small individuals; and (4) altitudinal variation in community body sizes is weak and probably obscured by other strong local environmental gradients such as precipitation.

Thus, Bergmann's rule appears to not apply as a general feature of land snail biogeography. The reasons for this are related to the unique physiology and biology of land snails: as they possess limited capabilities for lowering their freezing temperature and are not capable of active dispersal across landscapes, small forms will be favored at high latitudes, as they will maintain better supercooling abilities and will more rapidly and effectively colonize deglaciated landscapes.

This result cautions that general latitudinal body size predictions for ectotherms must consider more than heat balance issues if they are to reflect real world patterns. For instance, ectotherms not exposed to freezing temperatures, or with better abilities to generate cryoprotection compounds or to actively disperse, may exhibit different body size relationships from those seen in land snails. The prevalence of Bergmann-style body size variations in marine invertebrates (Poulin 1995b, 1995a; Poulin and Hamilton 1995; Poulin 1996) may be due to the fact that these populations are never exposed to freezing temperatures. Additionally, the colonial nature of ants may buffer individuals against winter temperatures, allowing this taxa group to follow predictions of Bergmann's rule (Kaspari and Vargo 1995). It is thus imperative that the unique physiology, reproductive biology, behavior, and environments of ectothermic groups be considered when formulating body size hypotheses.

## References

Ansart, A. 2002. "Hibernation et résistance au froid chez l'escargot petit-gris *Cornu aspersum* (syn. *Helix aspersa*; Gastéropode, Pulmoné)." *Bulletin de la Société zoologique de France* 127 (4): 375–380.

Ansart, A., and P. Vernon. 2003. "Cold hardiness in molluscs." *Acta Oecologica* 24 (2): 95–102.

———. 2004. "Cold hardiness abilities vary with the size of the land snail *Cornu aspersum*." *Comparative Biochemistry and Physiology A: Molecular and Integrative Physiology* 139 (2): 205–211. doi: 10.1016/J.Cbpb.2004.09.003.

Ansart, A., P. Vernon, and J. Daguzan. 2002. "Elements of cold hardiness in a littoral population of the land snail *Helix aspersa* (Gastropoda: Pulmonata)." *Journal of Comparative Physiology B: Biochemical, Systemic, and Environmental Physiology* 172 (7): 619–625.

Arad, Z., and T. R. Avivi. 1998. "Ontogeny of resistance to desiccation in the bush-dwelling snail *Theba pisana* (Helicidae)." *Journal of Zoology* 244:515–526.

Arad, Z., S. Goldenberg, and J. Heller. 1998. "Short- and long-term resistance to desiccation in a minute litter-dwelling land snail *Lauria cylindracea* (Pulmonata: Pupillidae)." *Journal of Zoology* 246 (1): 75–81. doi: 10.1111/j.1469-7998.1998.tb00134.x.

Armbruster, G. F. J. 2001. "Selection and habitat-specific allozyme variation in the self-fertilizing land snail *Cochlicopa lubrica* (O. F. Müller)." *Journal of Natural History* 35 (2): 185–199.

Ashton, K. G., M. C. Tracy, and A. de Queiroz. 2000. "Is Bergmann's rule valid for mammals?" *American Naturalist* 156 (4): 390–415.

Barker, G. M. 2005. "The character of the New Zealand land snail fauna and communities: Some evolutionary and ecological perspectives." *Records of the Western Australian Museum* 68:53–102.

Barlow, N. D. 1994. "Size distributions of butterfly species and the effect of latitude on species sizes." *Oikos* 71 (2): 326–332.

Baur, B. 1988. "Microgeographical variation in shell size of the land snail *Chondrina clienta*." *Biological Journal of the Linnean Society* 35 (3): 247–259.

Bequaert, J. C., and W. B. Miller. 1973. *The mollusks of the arid southwest: With an Arizona check list.* Tucson: University of Arizona Press.

Bergmann, C. 1847. "Ueber die Verhältnisse der Wärmeökonomie der Thiere zu ihrer Grösse." *Göttinger Studien* 3:595–708.

Blackburn, T. M., K. J. Gaston, and N. Loder. 1999. "Geographic gradients in body size: A clarification of Bergmann's rule." *Diversity and Distributions* 5:165–174.

Cameron, R. A. D., and B. M. Pokryszko. 2005. "Estimating the species richness and composition of land mollusc communities: Problems, consequences and practical advice." *Journal of Conchology* 38:529–547.

Cowie, R. H. 1995. "Variation in species diversity and shell shape in Hawaiian land snails: In situ speciation and ecological relationships." *Evolution* 49 (6): 1191–1202.

Cushman, J. H., J. H. Lawton, and B. F. J. Manly. 1993. "Latitudinal patterns in European ant assemblages: Variation in species richness and body size." *Oecologia* 95 (1): 30–37.

Dillon, T. J., and A. L. Metcalf. 1997. "Altitudinal distribution of land snails in some montane canyons in New Mexico." In *Land snails of New Mexico*, Bulletin of the New Mexico Museum of Natural History and Science 10, 109–127. Albuquerque: New Mexico Museum of Natural History and Science.

Emberton, K. C. 1995. Land-snail community morphologies of the highest-diversity sites of Madagascar, North America, and New Zealand, with recommended alternatives to height-diameter plots. *Malacologia* 36:43–66.

Foster, B. A., and A. E. Stiven. 1994. "Effects of age, body size, and site on reproduction in the southern Appalachian land snail *Mesodon normalis* (Pilsbry, 1900)." *American Midland Naturalist* 132 (2): 294–301.

——. 1996. "Experimental effects of density and food on growth and mortality of the southern Appalachian land gastropod, *Mesodon normalis* (Pilsbry)." *American Midland Naturalist* 136 (2): 300–314.

Goodfriend, G. A. 1986. "Variation in land-snail shell form and size in its causes: A review." *Systematic Zoology* 35 (2): 204–223.

Hausdorf, B. 2003. "Latitudinal and altitudinal body size variation among north-west European land snail species." *Global Ecology and Biogeography* 12 (5): 389–394.

Hausdorf, B., and C. Hennig. 2003. "Nestedness of north-west European land snail ranges as a consequence of differential immigration from Pleistocene glacial refuges." *Oecologia* 135 (1): 102–109.

Hawkins, B. A. 1995. "Latitudinal body-size gradients for the bees of the eastern United States." *Ecological Entomology* 20 (2): 195–198.

Hawkins, B. A., and J. H. Lawton. 1995. "Latitudinal gradients in butterfly body sizes: Is there a general pattern?" *Oecologia* 102 (1): 31–36.

Hayakaze, E., and S. Chiba. 1999. "Historical and ontogenetic changes in shell width and shape of land snails on the island of Kikai." *American Malacological Bulletin* 15 (1): 75–82.

Heller, J. 2001. "Life history strategies." In *The biology of terrestrial molluscs*, edited by G. M. Barker, 413–445. Wallingford, New Zealand: CABI Publishing.

Hubbell, S. P. 2001. *The unified neutral theory of biodiversity and biogeography.* Princeton: Princeton University Press.

Hubricht, L. 1985. "The distributions of the native land mollusks of the eastern United States." *Fieldiana* 24:1–191.

Hurlbert, A. H., and E. P. White. 2005. "Disparity between range map- and survey-based analyses of species richness: Patterns, processes and implications." *Ecology Letters* 8 (3): 319–327. doi: 10.1111/j.1461-0248.2005.00726.x.

James, F. C. 1970. "Geographic size variation in birds and its relationship to climate." *Ecology* 51 (3): 365–390.

Kaspari, M., and E. L. Vargo. 1995. "Colony size as a buffer against seasonality: Bergmann's rule in social insects." *American Naturalist* 145 (4): 610–632.

Kerney, M. P., R. A. D. Cameron, and J. H. Jungbluth. 1983. *Die Landschnecken Nord- und Mitteleuropas*. Hamburg: Parey.

Kirchner, C., R. Krätzner, and F. W. Welter-Schultes. 1997. "Flying snails: How far can *Truncatellina* (Pulmonata: Vertiginidae) be blown over the sea?" *Journal of Molluscan Studies* 63:479–487.

Lee, R. E., and J. P. Costanzo. 1998. "Biological ice nucleation and ice distribution in cold-hardy ectothermic animals." *Annual Review of Physiology* 60:55–72.

Madec, L., A. Bellido, and A. Guiller. 2003. "Shell shape of the land snail *Cornu aspersum* in North Africa: Unexpected evidence of a phylogeographical splitting." *Heredity* 91 (3): 224–231. doi: 10.1038/Sj.Hdy.6800301.

Marui, Y., S. Chiba, J. Okuno, and K. Yamasaki. 2004. "Species-area curve for land snails on Kikai Island in geological time." *Paleobiology* 30 (2): 222–230.

Mayr, E. 1956. "Geographical character gradients and climatic adaptation." *Evolution* (10): 105–108.

Metcalf, A. L., and R. A. Smartt. 1997. *Land snails of New Mexico*. Bulletin of the New Mexico Museum of Natural History and Science 10. Albuquerque: New Mexico Museum of Natural History and Science.

Nekola, J. C. 2005. "Geographic variation in richness and shell size of eastern North American land snail communities." *Records of the Western Australian Museum* 68:39–51.

Nekola, J., and M. Barthel. 2002. "Morphometric analysis of the genus *Carychium* in the Great Lakes region of North America." *Journal of Conchology* 37 (5): 515–531.

Nekola, J. C., and B. F. Coles. 2001. "Systematics and ecology of *Gastrocopta* (*Gastrocopta*) *rogersensis* (Gastropoda: Pupillidae), a new species of land snail from the Midwest of the United States of America." *Nautilus* 115: 105–114.

Pfenninger, M. 2004. "Comparative analysis of range sizes in *Helicidae* s.1. (Pulmonata, Gastropoda)." *Evolutionary Ecology Research* 6 (3): 359–376.

Pilsbry, H. A. 1948. *Land Mollusca of North America (North of Mexico)*. Academy of Natural Sciences of Philadelphia Monograph 3. Philadelphia: Academy of Natural Sciences of Philadelphia.

Pokryszko, B. M., and R. A. D. Cameron. 2005. "Geographical variation in the composition and richness of forest snail faunas in northern Europe." *Records of the Western Australian Museum* 68:115–132.

Popov, V. N., and S. S. Kramarenko. 2004. "Dispersal of land snails of the genus Xeropicta Monterosato, 1892 (Gastropoda; Pulmonata; Hygromiidae)." *Russian Journal of Ecology* 35 (4): 263–266.

Poulin, R. 1995a. "Clutch size and egg size in free-living and parasitic copepods: A comparative analysis." *Evolution* 49 (2): 325–336.

———. 1995b. "Evolutionary influences on body size in free-living and parasitic isopods." *Biological Journal of the Linnean Society* 54 (3): 231–244.

———. 1996. "The evolution of body size in the Monogenea: The role of host size and latitude." *Canadian Journal of Zoology* 74 (4): 726–732.

Poulin, R., and W. J. Hamilton. 1995. "Ecological determinants of body size and clutch size in amphipods: A comparative approach." *Functional Ecology* 9 (3): 364–370.

Rensch, B. 1938. "Some problems of geographical variation and species-formation." *Proceedings of the Linnean Society of London* 150:275–285.

Riddle, W. A. 1983. "Physiological ecology of land snails and slugs." In *The Mollusca*, vol. 6, *Ecology*, edited by W. D. Russell-Hunter, 431–461. New York: Academic Press.

Riddle, W. A., and V. J. Miller. 1988. "Cold-hardiness in several species of land snails." *Journal of Thermal Biology* 13 (4): 163–167.

Roy, K., and K. K. Martien. 2001. "Latitudinal distribution of body size in northeastern Pacific marine bivalves." *Journal of Biogeography* 28 (4): 485–493.

Schilthuizen, M., and M. Lombaerts. 1994. "Population structure and levels of gene flow in the Mediterranean land snail *Albinaria corrugata* (Pulmonata: Clausiliidae)." *Evolution* 48 (3): 577–586.

Stiven, A. E., and B. A. Foster. 1996. "Density and adult size in natural populations of a southern Appalachian low-density land snail, *Mesodon normalis* (Pilsbry)." *American Midland Naturalist* 136 (2): 287–299.

Tryjanowski, P., and E. Koralewska-Batura. 2000. "Inter-habitat shell morphometric differentiation of the snail *Helix lutescens* Rossm. (Gastropoda: Pulmonata)." *Ekologia (Bratislava)* 19 (1): 111–116.

Turner, J. R. G., and J. J. Lennon. 1989. "Species richness and the energy theory: Reply." *Nature* 340 (6232): 351–351.

Vagvolgyi, J. 1975. "Body size, aerial dispersal, and origin of the Pacific land snail fauna." *Systematic Zoology* 24 (4): 465–488.

Welter-Schultes, F. W. 2000. "The pattern of geographical and altitudinal variation in the land snail *Albinaria idaea* from Crete (Gastropoda: Clausiliidae)." *Biological Journal of the Linnean Society* 71 (2): 237–250.

———. 2001. "Spatial variations in *Albinaria terebra* land snail morphology in Crete (Pulmonata: Clausiliidae): Constraints for older and younger colonizations?" *Paleobiology* 27 (2): 348–368.

CHAPTER THREE

# Geographic Variation in Body Size Distributions of Continental Avifauna

Brian A. Maurer

The statistical distribution of body sizes of species of related organisms is determined by a variety of factors (Brown 1995; Maurer 1999; Gaston and Blackburn 2000; Roy et al. 2000). These factors include evolutionary history and constraints (Stanley 1973; McKinney 1990; Maurer 1998b, 2003), the properties of the geographic space containing the assemblage being measured (Maurer et al. 1992; Blackburn and Gaston 1996), and the scale at which species are accumulated (Brown and Nicoletto 1991). Since continents are effectively open systems with respect to birds, colonization is a regularly occurring phenomenon on all but the most isolated islands. The accumulation of species of different body sizes on a continent occurs both by speciation and by the addition of lineages that originated on another continent. Because of these complications, the evolutionary history of an assemblage of bird species on a continent cannot be reliably recovered using phylogenetic analyses alone (Purvis et al. 2003; Volger and Ribera 2003). Although the history of certain monophyletic lineages of birds on some continents may be amenable to historical reconstruction, there is no guarantee that the entire avifauna of a continent will have the same, or even a similar, evolutionary history. Hence, determining the contribution of phylogenetic processes to the distribution of body sizes of bird species cannot be done without examining the properties of continents that influence the processes of speciation, extinction, and colonization.

In this chapter, I examine variation in the body size distributions of birds on different continents. Complete phylogenies are not available for most groups of birds, and ordinal relationships among birds are currently under scrutiny. Therefore, as a proxy for evolutionary history, I examine the taxonomic structure of avian body size distributions to assess at what taxonomic level divergence has been most pronounced (Smith et al. 2004). Given an estimate of the taxonomic structure of body size distributions across different continents, I then examine the likelihood that the distributions of each continent are congruent with the cumulative distribution of body sizes of birds of the world. If each continental distribution approximates the distribution of all birds, then this implies that the properties of body size distributions are independent of scale, at least at the scale of continents. It is well known that within continents, there is a striking scale dependence of body size distributions (Brown and Maurer 1989; Brown and Nicoletto 1991), but does this scale dependence extend to global scales? It is not clear that it should (Smith et al. 2004). It is possible, however, that the body size distributions of continents might represent nonrandom "draws" from the statistical distribution of body sizes for all birds. If this is true, the implication is that each continent poses unique or different ecological and/or historical conditions that "filter" the range of body sizes of birds to a distinct subset of species. Therefore, it is important to examine attributes of continents that might influence the distribution of body sizes within each continent.

Here I consider the relationship between continent size and properties of avian body size distributions. Continent size may act as a filter that eliminates species of certain size from maintaining viable populations on a continent (Brown and Maurer 1987, 1989; Maurer et al. 1992). If this is true, then body size distributions would be nonrandom subsets of the global distribution of bird body sizes. Furthermore, body size distributions should vary in a predictable fashion with continent size.

## Data and Methods

Data on the body sizes of birds were obtained from Dunning (1993) and processed as described by Maurer et al. (2004). Each species in this database was located in one of seven major geographic regions (table 3.1). These regions were defined purely on a geographic basis rather than reorganized into specific biogeographic regions, since there is a variety of ways to identify such regions (Brown and Lomolino 1998). For each geo-

TABLE 3.1 **Geographic Regions Used for Analyzing Geographic Patterns in Body Size Distributions of Birds**

| Geographic Region | Approximate Size ($10^5$ km$^2$) | Land Masses Included |
|---|---|---|
| Africa | 292.4 | Africa |
| Australia | 85.2 | Australia, New Guinea, New Zealand, Tasmania |
| Eurasia | 513.4 | Asia, Europe, Japan, Taiwan |
| Madagascar | 5.9 | Madagascar |
| North America | 242.1 | North America, Central America, Caribbean islands |
| South America | 177.7 | South America |
| South Pacific | 25.8 | Indonesia, Malaysia, Philippines, South Pacific islands |

graphic region, a list of species from the overall database was obtained by using locations for each species given by Howard and Moore (1991). A species was included in all geographic regions in which it occurred. Of the 6,430 species in the database, 262 were found in more than one geographic region. The taxonomy used was that of Sibley and Monroe (Sibley and Monroe 1990; Monroe and Sibley 1993; Sibley and Monroe 1993). The taxonomic ranks used were order, family, genus, and species.

For statistical analyses, all body sizes were transformed to common logarithms (base 10). Frequency histograms were constructed for the entire sample of species and for individual geographic regions using a standard set of body size intervals (each quarter logarithm from zero to five). To examine the taxonomic structure of each geographic region and for birds as a group, a nested analysis of variance was performed (Harvey and Pagel 1991; Smith et al. 2004). The variance components estimated from these analyses represent the degree to which variation in body size among birds in a geographic region can be explained by deep evolutionary history (orders) as opposed to recent history (e.g., genera). Significance tests for variance components were not available due to the unbalanced nature of the design. The degree to which taxonomy is a proxy for an accurate estimate of evolutionary history is determined by many factors. Although there are undoubtedly many reasons why current taxonomic knowledge regarding birds is incomplete, it is unlikely that the major patterns described below will be markedly altered by future refinements.

To examine the distinctiveness of the log body size distribution within each geographic region, the accumulated distribution of body sizes for all species across all continents was used as the "null" distribution against which distributions for individual regions were compared. Empirical likelihood functions (ELFs) were constructed for each frequency distribution (Maurer et al. 2004). These functions give the likelihood that a statistical moment (mean, variance, skewness, kurtosis) is derived from a specified frequency distribution. Comparing the likelihood that the mean, say, for a geographic region was derived from the distribution of all species to the same likelihood derived from the distribution for that region gives a relative measure of the evidence that the two distributions are identical. The negative of the logarithm of the ratio of these two likelihoods increases as the distributions are less similar to one another (Maurer et al. 2004). Although the sampling distribution for these empirical likelihood ratios (ELRs) is unknown, large ELRs (> ~8.0) provide evidence against the supposition that the moments come from two identical distributions (Edwards 1992; Royall 1997, 2004).

Continent size was determined by calculating the total area of terrestrial land masses in the geographic regions described in table 3.1. These areas were taken from projections available in the ArcGIS program. Simple linear regressions of each of the four statistical moments from log body size distributions on the common logarithm of continent size were performed.

## The Patterns

As has been shown elsewhere (Blackburn and Gaston 1994; Maurer 1998a), the global distribution of log body sizes for birds is positively skewed (fig 3.1, table 3.2). The average global log body size corresponds to a mass of about 55 g, and the standard deviation of the distribution is approximately 5 g. The global distribution is also slightly leptokurtic (peaked). The means of the continental distributions were all substantially different from the mean of the global distribution (table 3.2). South America had a smaller mean, while all the other continents had larger means. There was no evidence in the data that variances of African, Australian, and North American distributions were different from the global distribution (fig. 3.2, table 3.2). Eurasia had a greater variance than the global variance (fig. 3.2, table 3.2), while South America, Madagascar, and the South Pacific region had smaller variances (figs. 3.2 and 3.3, table 3.2). All regional distributions were positively skewed (figs. 3.2 and 3.3, table 3.2). Distributions for all continental regions except for South America were less skewed than the global value (table 3.2). Only

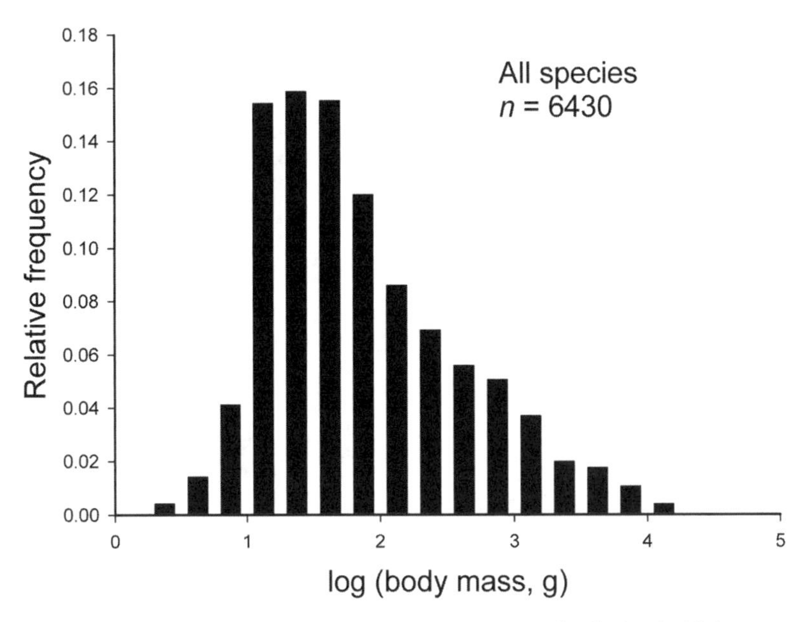


FIGURE 3.1. Frequency histogram of the global log body mass distribution for birds

TABLE 3.2 **Estimates of the First Four Statistical Moments of Log Body Size Distributions for All Birds of The World, and for the Geographic Regions Defined in Table 3.1**

| Geographic Region | Mean | Variance | Skewness | Kurtosis |
|---|---|---|---|---|
| All regions | 1.74 | 0.53 | 0.82 | 0.21 |
| Africa | **1.90** | 0.54 | **0.72** | **−0.15** |
| Australia | **1.89** | 0.53 | **0.55** | **−0.33** |
| Eurasia | **1.91** | **0.61** | **0.59** | **−0.55** |
| Madagascar | **1.91** | **0.40** | **0.21** | **−0.81** |
| North America | **1.78** | 0.55 | **0.49** | **−0.35** |
| South America | **1.65** | **0.49** | 0.76 | 0.19 |
| South Pacific | **1.85** | **0.46** | **0.42** | **−0.46** |

*Note*: Estimates in bold indicate that the likelihood ratio comparing the continental value to the global value was greater than 8. Such ratios indicate that the data contained evidence for a difference of the regional value from the global value.

South America had a leptokurtic distribution that was similar to the global value (table 3.2). All the other regional distributions were flatter (more platykurtic) than the global value (table 3.2).

The taxonomic structure of the global log body size distribution was dominated by variation among orders (table 3.3). Variation among families explained the next-greatest variation, followed by variation among genera. This pattern was not seen within individual geographic regions,

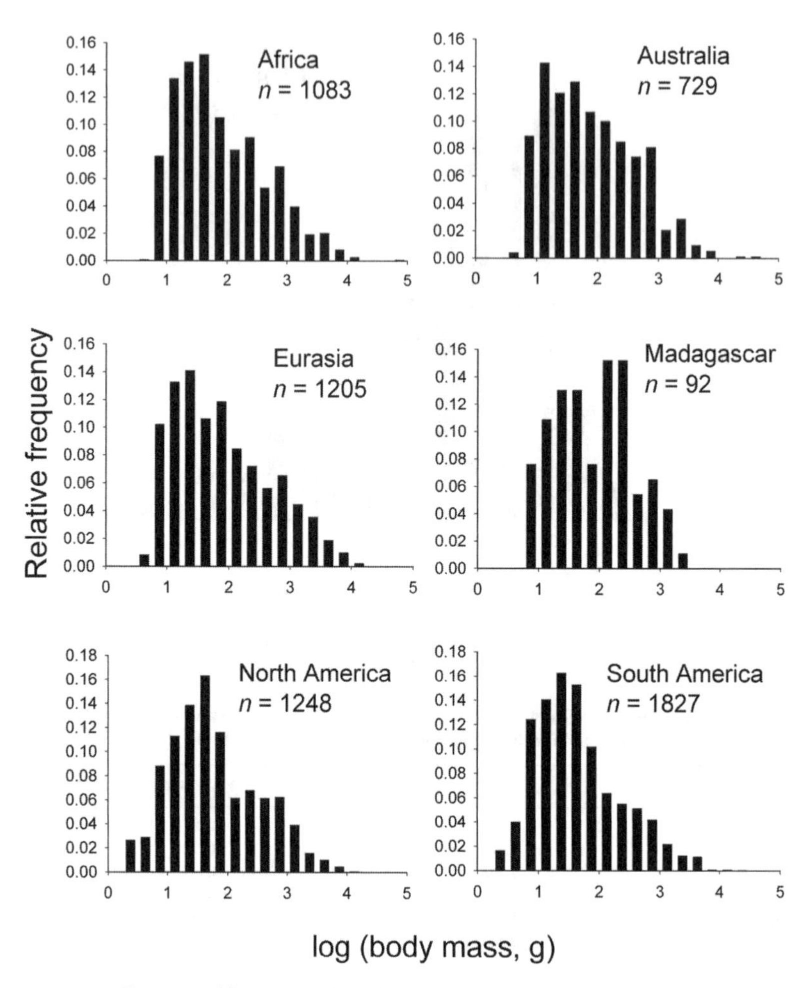


FIGURE 3.2. Frequency histograms of continental log body mass distribution for birds

however. Although ordinal variation explained roughly two-thirds of the variation in log body sizes in all regions, in no region was variation among families greater than variation among genera (table 3.3). Madagascar was the most distinctive region, with over three-quarters of the variance explained by orders and only 2% explained by variation among families.

There was no relationship between average log body size and continent size (fig. 3.4). However, there was a strong positive relationship between the variance and continent size, indicating that larger continents had more

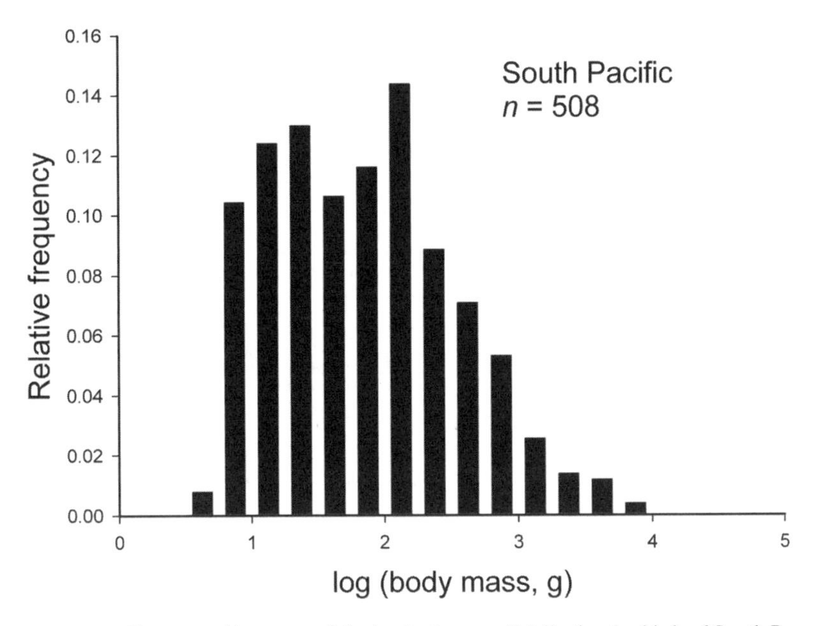


FIGURE 3.3. Frequency histogram of the log body mass distribution for birds of South Pacific islands

TABLE 3.3 **Nested Analyses of Variance for Log Body Masses of All Birds and of Birds within Each Geographic Region Defined in Table 3.1**

| Geographic Region | Taxon | Sample Size | Variance Component | % of Total Variance |
|---|---|---|---|---|
| All regions | Order | 31 | 0.46 | 67.3 |
| | Family | 160 | 0.11 | 15.9 |
| | Genus | 1,567 | 0.08 | 12.0 |
| Africa | Order | 26 | 0.44 | 70.8 |
| | Family | 64 | 0.08 | 12.3 |
| | Genus | 238 | 0.08 | 12.9 |
| Australia | Order | 22 | 0.44 | 68.9 |
| | Family | 64 | 0.08 | 12.5 |
| | Genus | 201 | 0.09 | 14.7 |
| Eurasia | Order | 23 | 0.57 | 74.9 |
| | Family | 68 | 0.07 | 8.9 |
| | Genus | 292 | 0.10 | 12.8 |
| Madagascar | Order | 17 | 0.34 | 76.2 |
| | Family | 27 | 0.01 | 2.7 |
| | Genus | 33 | 0.06 | 13.6 |
| North America | Order | 24 | 0.52 | 77.5 |
| | Family | 60 | 0.06 | 8.7 |
| | Genus | 445 | 0.07 | 10.8 |
| South America | Order | 25 | 0.48 | 75.6 |
| | Family | 63 | 0.06 | 10.0 |
| | Genus | 596 | 0.08 | 11.8 |
| South Pacific | Order | 20 | 0.35 | 65.5 |
| | Family | 56 | 0.06 | 11.1 |
| | Genus | 167 | 0.10 | 19.0 |

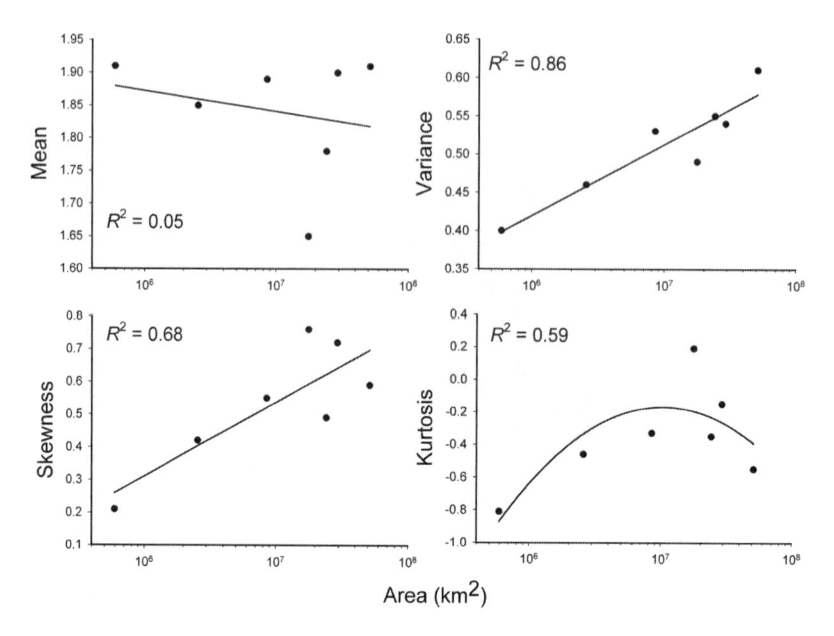


FIGURE 3.4. Relationships of moments of log body mass distributions to the areas of seven geographic regions defined in table 3.1

variability among species in their body sizes. There was also a positive relationship between the continent area and the skewness of the log body size distribution. That is, the larger the continent, the more asymmetrical and positive the skew of the log body size distribution. The relationship between kurtosis and continental area was nonlinear (fig. 3.4). Distributions became more peaked (higher kurtosis) with area up to South America, after which the distributions for the larger continents became flatter, indicating a higher density of species in the tails of those distributions.

## Discussion

The data on body sizes for birds indicate that the structure of body size distributions is scale dependent and is influenced by the size of continents. Distributions for continents and geographic regions are not simple random mixtures of species from the global distribution but instead are "filtered" by continent size in at least two ways. First, larger continents have a wider range of body sizes than smaller continents, indicated by the increase in variance in log body size distributions with larger continents. Second, larger continents tend to have a disproportionate share of large species. That is, there is more probability in the right tail of log body size distributions on larger continents, resulting in a more skewed distribution.

What is the biogeographic mechanism that is responsible for this filtering of body size as a function of continent size? There is a minimum size to the geographic range of a species that increases with body size (Brown 1981; Brown and Maurer 1987). Larger species must occur at lower densities (Damuth 1981; Brown and Maurer 1987; Damuth 1987, 1991) because they must extract energy from a larger area than smaller species (Brown 1995). To maintain a viable population, the minimum area inhabited by a large species must compensate for its lower density. Hence, species of large size with small ranges will have a greater chance of extinction due to the lower total population size achievable in a small range (Brown and Maurer 1987, 1989). Since small continents limit the size of geographic ranges, on the average, they should have relatively fewer large species than larger continents.

It is important to briefly consider the nature of the processes that lead to this filtering of body sizes as a function of continent size. Clearly, if range size is involved, then extinction of propagules of large organisms on small continents may play an important role in maintaining the properties of body mass distributions on small continents. However, it is also possible that species of large organisms that invade small continents experience intense natural selection toward decreasing size, resulting in dwarfism. Indeed, evolution of small size on islands has been seen repeatedly in mammals (Lomolino 1985; Marquet and Taper 1998). A similar phenomenon has yet to be documented for birds. It is less clear how speciation might vary among large and small species across continents. Presumably, the same factors that might discourage establishment of large species that disperse to small continents may prevent successful speciation of large forms on smaller continents.

It is possible that the higher variability and skewness of log body mass distributions for larger continents might be partially related to the fact that two of the largest three continents are the furthest north of all the continents and therefore have a larger proportion of more severe cold habitats (e.g., tundra) than the other continents. Presumably it is physiologically advantageous to be a larger warm-blooded vertebrate in a colder climate. However, although North America has a larger variance than either Africa or South America, both of the latter continents have higher positive skewness (table 3.2). One would expect a continent with more cold habitats to be more skewed toward large body sizes.

Interestingly, variance components analysis indicated that variation in log body masses among genera was equal to or exceeded variation among families on continents, although the opposite was true for the global dataset. In mammals, continental log body mass distributions had familial variance components that exceeded generic variance components for four continents for which data were available (Smith et al. 2004). To the degree that taxonomy of birds reflects evolutionary history in a manner similar to taxonomy of mammals, this result implies that continental variation in body size in birds is attributable to more recent evolutionary events than that of mammals. This, in turn, implies a greater "mixing" of bird taxa over the recent geological past than of mammal taxa. This is exactly what might be expected for a more vagile group of organisms such as birds. On the average, geographic barriers may be more easily crossed by birds than mammals. Of course, the difference between variance partitioning in birds and mammals might be attributable to differences in the way that taxonomists recognize genera in these two groups. For example, if ornithologists tended to be more liberal in recognizing criteria as evidence for the existence of a new genus than mammalogists, the higher variance components for bird genera could be artifactual. Unfortunately, adequate phylogenetic information is not available for a wide enough variety of taxa for both birds and mammals to resolve this issue.

To date, there have been few analyses of body size variation of large groups of organisms across global geographic gradients. It is therefore premature to attempt to generalize too much regarding how the shapes of body mass distributions are affected by these gradients. If these results for birds are applicable to other groups, we might expect to see scale dependence in continental body size distributions of other groups. Clearly there is a need for more data from a variety of taxa before such patterns can be examined.

## References

Blackburn, T. M., and K. J. Gaston. 1994. "The distribution of body sizes of the world's bird species." *Oikos* 70 (1): 127–130.

———. 1996. "Spatial patterns in the body sizes of bird species in the New World." *Oikos* 77: 436–446.

Brown, J. H. 1981. "Two decades of homage to Santa Rosalia: Toward a general theory of diversity." *American Zoologist* 21: 877–888.

———. 1995. *Macroecology*. Chicago: University of Chicago Press.

Brown, J. H., and M. V. Lomolino. 1998. *Biogeography*. 2nd ed. Sunderland, MA: Sinauer Associates.

Brown, J. H., and B. A. Maurer. 1987. "Evolution of species assemblages: Effects of energetic constraints and species dynamics on the diversification of the North American avifauna." *American Naturalist* 130 (1): 1–17.

———. 1989. "Macroecology: The division of food and space among species on continents." *Science* 243 (4895): 1145–1150.

Brown, J. H., and P. F. Nicoletto. 1991. "Spatial scaling of species composition: Body masses of North American land mammals." *American Naturalist* 138 (6): 1478–1512.

Damuth, J. D. 1981. "Population density and body size in mammals." *Nature* 290: 699–700.

———. 1987. "Interspecific allometry of population density in mammals and other animals: The independence of body mass and population energy-use." *Biological Journal of the Linnean Society* 31:193–246.

———. 1991. "Of size and abundance." *Nature* 351:268–269.

Dunning, J. B., Jr. 1993. *CRC handbook of avian body masses*. Boca Raton, FL: CRC Press.

Edwards, A. W. F. 1992. *Likelihood*. Baltimore: Johns Hopkins University Press.

Gaston, K. J., and T. M. Blackburn. 2000. *Pattern and process in macroecology*. Oxford: Blackwell Science.

Harvey, P. H., and M. D. Pagel. 1991. *The comparative method in evolutionary biology*. Oxford: Oxford University Press.

Howard, R., and A. Moore. 1991. *A complete checklist of the birds of the world*. 2nd ed. London: Academic Press.

Lomolino, M. V. 1985. "Body size of mammals on islands: The island rule reexamined." *American Naturalist* 125 (2): 310–316.

Marquet, P. A., and M. L. Taper. 1998. "On size and area: Patterns of mammalian body size extremes across landmasses." *Evolutionary Ecology* 12 (2): 127–139.

Maurer, B. A. 1998a. "The evolution of body size in birds. I. evidence for nonrandom diversification." *Evolutionary Ecology* 12:925–934.

———. 1998b. "The evolution of body size in birds. II. The role of reproductive power." *Evolutionary Ecology* 12:935–944.

———. 1999. *Untangling ecological complexity: The macroscopic perspective*. Chicago: University of Chicago Press.

——. 2003. "Adaptive diversification of body size: The roles of physical constraint, energetics, and natural selection." In *Macroecology: Causes and consequences*, edited by T. M. Blackburn and K. J. Gaston, 174–191. Oxford: Blackwell.

Maurer, B. A., J. H. Brown, T. Dayan, B. J. Enquist, S. K. M. Ernest, E. A. Hadly, J. P. Haskell, D. Jablonski, K. E. Jones, D. M. Kaufman, S. K. Lyons, K. J. Niklas, W. P. Porter, K. Roy, F. A. Smith, B. Tiffney, and M. R. Willig. 2004. "Similarities in body size distributions of small-bodied flying vertebrates." *Evolutionary Ecology Research* 6 (6): 783–797.

Maurer, B. A., J. H. Brown, and R. D. Rusler. 1992. "The micro and macro in body size evolution." *Evolution* 46 (4): 939–953.

McKinney, M. L. 1990. "Trends in body size evolution." In *Evolutionary trends*, edited by K. J. McNamara, 75–118. Tucson: University of Arizona Press.

Monroe, B. L., and C. G. Sibley. 1993. *A world checklist of birds*. New Haven: Yale University Press.

Purvis, A., C. D. L. Orme, and K. Dolphin. 2003. "Why are most species small-bodied? A phylogenetic view." In *Macroecology: Concepts and consequences*, edited by T. M. Blackburn and K. J. Gaston, 155–173. Oxford: Blackwell.

Roy, K., D. Jablonski, and K. K. Martien. 2000. "Invariant size-frequency distributions along a latitudinal gradient in marine bivalves." *Proceedings of the National Academy of Sciences of the United States of America* 97 (24): 13150–13155.

Royall, R. M. 1997. *Statistical evidence: A likelihood paradigm*. London: Chapman and Hall.

——. 2004. "The likelihood paradigm for statistical evidence." In *The nature of scientific evidence*, edited by M. L. Taper and Subhash Lele, 119–152. Chicago: University of Chicago Press.

Sibley,C. G., and B. L. Monroe. 1990. *Distribution and taxonomy of birds of the world*. New Haven: Yale University Press.

——. 1993. *A supplement to "Distribution and taxonomy of birds of the world."* New Haven: Yale University Press.

Smith, F. A., J. H. Brown, J. P. Haskell, S. K. Lyons, J. Alroy, E. L. Charnov, T. Dayan, B. J. Enquist, S. K. M. Ernest, E. A. Hadly, K. E. Jones, D. M. Kaufman, P. A. Marquet, B. A. Maurer, K. J. Niklas, W. P. Porter, B. Tiffney, and M. R. Willig. 2004. "Similarity of mammalian body size across the taxonomic hierarchy and across space and time." *American Naturalist* 163 (5): 672–691.

Stanley, S. M. 1973. "An explanation for Cope's rule." *Evolution* 27:1–26.

Volger, A. P., and I. Ribera. 2003. "Evolutionary analysis of species richness patterns in aquatic beetles: Why macroecology needs a historical perspective." In *Macroecology: Concepts and consequences*, edited by T. M. Blackburn and K. J. Gaston, 17–30. Oxford: Blackwell.

CHAPTER FOUR

# Evolution of Body Size in Bats

Kamran Safi, Shai Meiri, and Kate E. Jones

Body mass, because of its central role in influencing morphology, physiology, ecology, and diversification of organisms (Stanley 1973; Cardillo et al. 2005; Olden et al. 2007), has been the focus of many evolutionary studies over the past two decades. Bats are likely to occupy a unique place within mammals due to the constraints imposed by aerodynamics and nocturnality, roosting posture, and echolocation (Norberg 1987; Barclay and Brigham 1991; Rayner 1996). These form unique selective pressures on the evolution of bat mass, as well as on bats' morphologies, life histories, and ecologies. Unfortunately despite the uniqueness of bats, they have been routinely omitted from macroecological and macroevolutionary studies of body mass. Bats often do not support predictions of evolutionary and ecological theories that were developed for other mammals or birds. For example, optimal body masses (Jones and Purvis 1997), brain size evolution (Jones and MacLarnon 2004), life history evolution (Jones and MacLarnon 2001), and geographic range and body size relationships (Willig et al. 2003) of bats differ from those of terrestrial mammals. Investigating the evolution of body mass in bats may thus provide a unique insight into body mass evolution.

In this chapter we use our comprehensive datasets of mammalian mass, ecology, life history, and spatial distribution (Grenyer et al. 2006; Jones et al. 2009) and our species-level phylogeny (Bininda-Emonds et al. 2007) coupled with results from previous analyses to provide an overview and a synthesis of the macroevolution and macroecology of bat body mass. Using body mass as a measure of size in bats has been criticized because individual body mass can triple in one evening after a foraging bout (Neuweiler 2000). The length of the forearm is often used as a more robust measure of size in bats. However, forearm length is a functionally important morphological feature of bat wings and is shaped by the demands of flight and under evolutionary selection pressures (Norberg and Rayner 1987; Norberg 1994). As the correlation ($r^2$) of the linear regression of ln (forearm lengths) to ln (body mass) in bats within the database is high (0.88) (Jones et al. 2009), we used forearm length to extrapolate body mass for species for which body mass measures were missing in all the following analyses. All our statistical analyses were carried out in R version 2.9.0 (R Development Core Team 2008). We follow the taxonomy of Wilson and Reeder (1993) to ensure consistency between data sources.

## Bat Body Mass Variation

The range of body mass in bats extends from the largest species weighing just over 1 kg (golden-capped fruit bat, *Acerodon jubatus*) to one of the smallest (bumblebee bat, *Craseonycteris thonglongyai*) representing one of the smallest known mammals at 2 g. The frequency distribution of bat body mass follows the trends in terrestrial mammals in that it is right skewed on a logarithmic scale (a few large species and many small species) (Brown and Nicoletto 1991; Rodriguez et al. 2008; Clauset et al. 2009) (fig. 4.1). Bats occupy a narrow subset of the masses within mammals (most observations fall within 6–16 g size classes) (fig. 4.1, *inset*) and, rather uniquely, have a bimodal size distribution. The lower mode at large body size may reflect the radiation of flying foxes (Pteropodidae) in the Old World tropics. A previous analysis with this dataset showed no significant sexual dimorphism in bats, although females are more often larger than males, which is unusual within mammals as a whole (Ralls 1976; Williams and Findley 1979; Lindenfors et al. 2007).

## Phylogenetic Signal in Body Mass

The amount of variation in traits that is due to shared evolutionary history, or "phylogenetic signal," can be explicitly calculated and is often the first step in understanding the mode and tempo of evolution of a particular trait. Some traits have been found to be more constrained by phylogenetic history than others; for example, behavioral traits are often

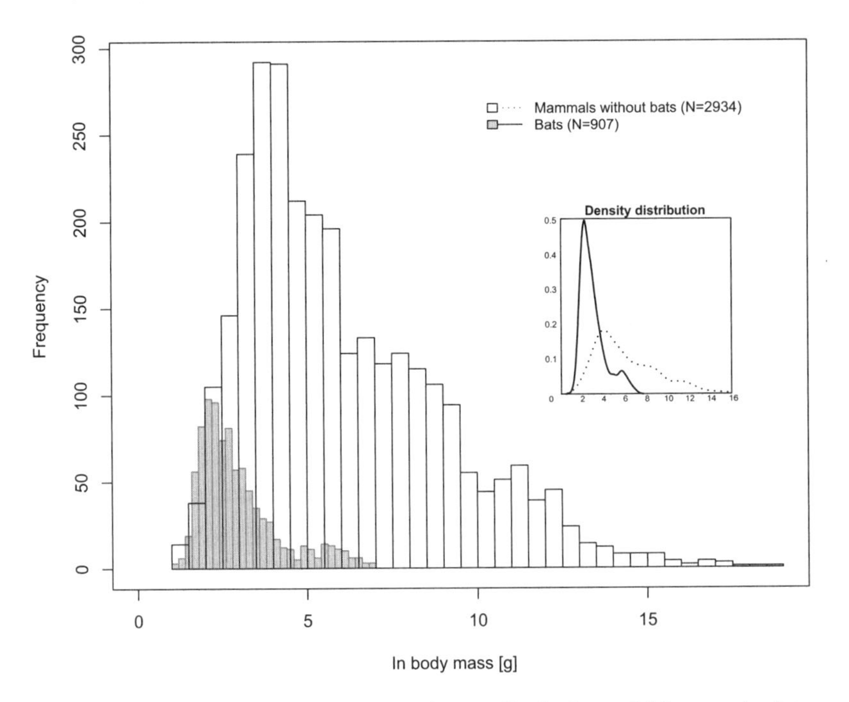


FIGURE 4.1. The frequency distribution of ln (mammalian body mass) (g) comparing bats to other mammals for 907 bat species and 2,934 terrestrial mammal species. *Inset*, density distribution showing the proportional distribution of body mass classes.

less phylogenetically constrained than size traits (Blomberg et al. 2003). There are a growing number of methods to estimate phylogenetic signal in traits. Here we use two to investigate body mass phylogenetic signal in bats: Moran's *I* (Gittleman and Kot 1990) and the parameter lambda ($\lambda$) (Pagel 1999). These methods try to partition the trait variance into the similarity due to shared ancestry and that caused by the selection that has occurred since divergence from a common ancestor.

Bats as well as the other mammalian orders and other taxa show a consistently high level of phylogenetic signal in body mass across these different methods (table 4.1; Smith et al. 2004). Values of Moran's *I* vary from +1 to −1, where positive values indicate that a trait at a particular phylogenetic level in the evolutionary tree is more similar than would be expected by chance, negative values indicate that traits are more different than expected, and zero values indicate that traits are distributed randomly with respect to phylogenetic distance across the tree. Bats have significantly positive *I* values across the entire phylogeny ($I$ = 0.25), cal-

TABLE 4.1 **Phylogenetic Signal in ln (Body Mass) among Selected Mammalian Clades**

| Clade | N | Moran's *I* (SD) | Lambda (SD) |
|---|---|---|---|
| Bats | 897 | 0.25 (0.005) | 0.91 (0.01) |
| Rodents | 1,428 | 0.11 (<0.001) | 0.96 (0.03) |
| Carnivores | 263 | 0.16 (0.006) | 0.99 (0.05) |
| Primates | 231 | 0.24 (0.007) | 1.0 (0.02) |

culated overall using a phylogenetic distance matrix as a measure of similarity. This means that closely related species have significantly more similar body masses than would be expected by chance. Similar results are found for lambda ($\lambda$ = 0.91 for bat ln (body mass)). Lambda values vary between zero and 1, where zero indicates that a trait has evolved among species with no respect to the phylogenetic relatedness in the specified tree (i.e., as if species were independent and unrelated to each other). Lambda is a more powerful test of phylogenetic signal, as it specifically uses the topology of the phylogeny instead using phylogenetic distances as a measure of phylogenetic similarity.

Bats, like other mammalian clades, have a high degree of phylogenetic signal in body mass. This suggests that in bats, and in mammals more generally, only a small amount of the variance in mass overall can be explained by adaptive selection occurring after the origination from a common ancestor.

## Mode and Tempo of Bat Body Mass Evolution

### *Mode of Body Mass Evolution*

We show above that there is a strong phylogenetic signal within bat body masses, but how has this phylogenetic signal changed over bat evolutionary history? Can we investigate which particular mode of trait evolution has been important? For example, if body mass has evolved through a Brownian motion or a "random walk," then we would expect the similarity in bat body masses to decrease in relation to the phylogenetic distance between species (i.e., more distantly related species could have very different body sizes). Recent analytical developments mean that the mode of evolution in traits can be directly quantified (e.g., Gillman 2007; Clauset and Erwin 2008; Mattila and Bokma 2008; Cooper and Purvis 2010). Statistical analyses of evolutionary modes focus on testing among three models: (1) Brownian motion (BR), (2) Ornstein-Uhlenbeck (OU), and (3) early-burst (EB). The BR model assumes gradual evolution of body mass over evolutionary time, where more closely related species are more similar and the variance in body mass increases with time (Felsenstein 1973). The OU model is a modified BR model where body mass is constantly pulled back toward an optimum by some kind of ecophysiological constraint (Hansen 1997). The EB model predicts that trait diversification is most rapid early in a lineage, with decreasing rates toward the present (Blomberg et al. 2003). The mode of body mass evolution was recently examined in mammals (Cooper and Purvis 2010) using this mammalian database (Jones et al. 2009). Cooper and Purvis (2010) found evidence to suggest that the OU model is the most likely model to explain the mode of body mass evolution in bats (compared to EB in mammals overall). One hypothesis for the difference in the mode of body mass evolution in bats compared to other mammals is that the limitations imposed by flight may have constrained the size evolution of bats more than other mammals (Cooper and Purvis 2010).

### *Reconstructing Ancestral Body Mass*

Using the distribution of body mass of current species and the pattern of their evolutionary relationships, it is possible to estimate trait ancestral values (e.g., Pagel 1999). This technique was used to estimate bat ancestral body mass assuming a random walk model of evolution (Safi and Dechmann 2005). The ancestor of modern bat species was estimated to have a body mass of around 19 g (Safi and Dechmann 2005), which is close to the median for all modern bats (fig. 4.1) and also matches the current estimates from the fossil records (Simmons and Conway 2003). This result suggests that selective forces determining bat body mass have been bidirectional. However, the accuracy of ancestral reconstruction is very dependent on the accuracy of the estimated evolutionary relationships among species and the assumed mode of evolution (Webster and Purvis 2002). Our understanding of evolutionary relationships in bats has been in flux over the past decade (see "Understanding Evolutionary Relationships among Bats" below), and we are only just beginning to understand different modes of trait evolution. Consequently, ancestral estimates such as these should be interpreted with caution (Webster and Purvis 2001, 2002).

### *Species Diversification Rates*

It has been hypothesized that body mass influences species diversification rates through decreasing extinction rates or increasing speciation rates (reviewed in Purvis et al. 2003). This hypothesis was fueled by the observation that there are more species at smaller sizes (Dial and Marzluff 1988) However, often these studies were confounded by the effect of evolutionary history, and there is little evidence that body mass is a significant correlate of diversification when methods which correct for the role of phylogeny are used (Orme et al. 2002), and this is also true in bats (Isaac et al. 2005). What has influenced the diversification rate of bats remains an extremely interesting unanswered question, but body mass does not seem to be an important contributor on its own.

## Understanding Evolutionary Relationships among Bats

An accepted, stable hypothesis about the evolutionary relationships of the clade of interest is essential for evolutionary analyses such as we present here. Unfortunately, our understanding of bat phylogenetic relationships has been far from stable until the past few years, which has hampered understanding. Controversy started in 1986. Pettigrew (1986) discovered that the system of neural connections between the midbrain and the retina of pteropodids (Pteropodidae, formerly classified in the suborder Megachiroptera; see below) was similar to that of primates, whereas other bats (the former suborder Microchiroptera) showed the putatively "primitive" system of connections, in common with all other mammals. This led to two decades of debate about whether bats were diphyletic with respect to primates. Although there is now overwhelming support for the monophyly of bats (e.g., Teeling et al. 2000; Murphy et al. 2001; Teeling et al. 2005), some authors are continuing the debate (Pettigrew 2008).

Even if we accept the monophyly of bats, interrelationships within the nineteen currently recognized extant and seven extinct bat families have also been debated. Bats had traditionally been placed in two monophyletic suborders, the Microchiroptera and the Megachiroptera (Dobson 1875) (fig. 4.2). However, the monophyly of the Microchiroptera is now disputed, as molecular evidence suggests that several of the microchiropteran families are more closely related to Pteropodidae (Megachiroptera) than to the remaining families. The most widely used division of the order places the Pteropodidae, Rhinolophidae, Hipposideridae, Megadermatidae, Craseonycteridae, and Rhinopomatidae in a new suborder, initially termed the Yinpterochiroptera (Springer et al. 2001), and leaves the remaining families in the suborder Yangochiroptera (fig. 4.2). Yangochiroptera was originally named by Koopman (1984), although at that time it excluded Nycteridae and Emballonuridae (Springer et al. 2001; Gunnell and Simmons 2005). These clades are now commonly referred to as Pteropodiformes and Vespertilioniformes for Yinpterochiroptera and Yangochiroptera, respectively (following Hutcheon and Kirsch 2006). This reevaluation of bat evolutionary relationships has important implications for our understanding of the evolution of their traits. For example, echolocation was traditionally thought to have evolved only once in the ancestors of microbats (Maltby et al. 2010). However, with the understanding that microbats are paraphyletic has come the realization that the evolution of echolocation must have been far more complicated,

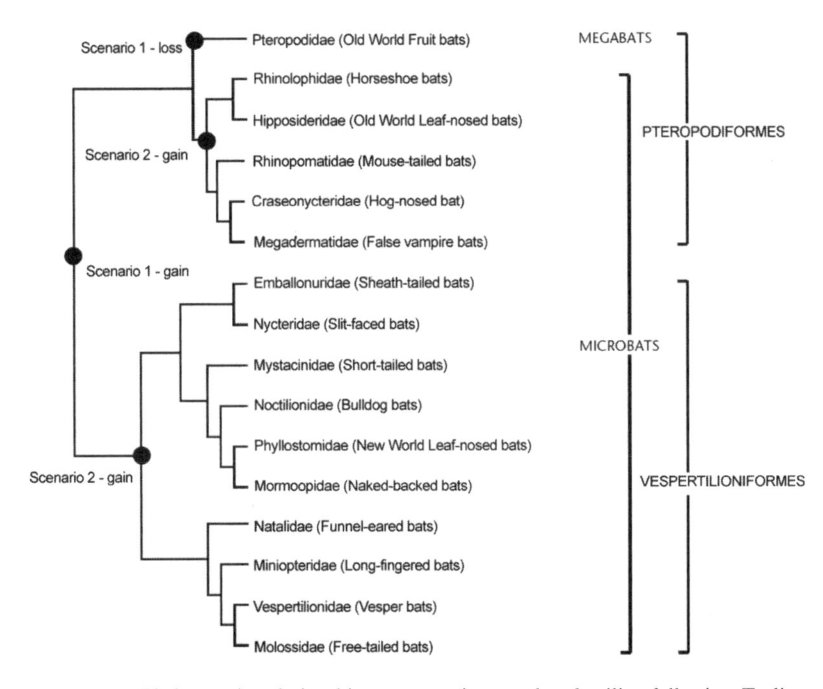


FIGURE 4.2. Phylogenetic relationships among nineteen bat families following Teeling et al.(2005). Circles indicate where echolocation may have evolved, either once (scenario 1) or twice (scenario 2), following Maltby et al. (2010).

either being gained and then lost (fig. 4.2, scenario 1) or gained twice independently (fig. 4.2, scenario 2). As our understanding of bat evolutionary relationships grows, there are undoubtedly many more interesting evolutionary patterns to be discovered.

## Understanding Body Mass Variation

Body mass is central to many species traits. For example, there is a tight correlation between body mass and many morphological, ecological, geographic, and life history traits (Stearns 1992; Calder 1996). However, what determines body mass?

### *Biological Correlates*

The documented tight relationship of body mass and many life history traits has led to the development of optimal life history theories (e.g., Charnov 1991, 1993; Kozlowski and Weiner 1997). In these approaches, one key life history stage is optimized, and body mass is the result of this trade-off. For example, according to Charnov's (1991) model, selection acts to maximize an individual's lifetime reproductive success given the mortality imposed by the environment. Altering the age at which sexual maturity is reached enables the optimization of reproductive success and therefore determines the size at which you mature and stop growing. How do bats fit with these models? Bat life histories do not have such a tight relationship with body mass compared to other mammals (Purvis and Harvey 1995; Jones and Purvis 1997), and key predictions from Charnov's (1991) model are not met (Jones and MacLarnon 2001). One of the reasons that bats do not fit may be that flight constrains bats to reach adult size before reaching reproductive maturity (near-adult dimensions are necessary for flight to occur). This factor of bat biology, unique among mammals, may impose unique constraints on life history features. This evidence suggests that evolution of life history traits in bats and body mass have been decoupled to some extent, leading to such observations as that they are the longest-lived mammal for their body mass (Jones and MacLarnon 2001). Research into explaining the contrary patterns in bats seems fruitful for further understanding the evolution of body mass.

*Environmental Correlates*

The variation in the spatial distribution of mammalian body mass has led to the development of theories which link body mass to environmental variables and other proxies such as latitude, for example, Bergmann's rule (Bergmann 1847; Mayr 1956). Bergmann's rule suggests that mammalian size increases with increasing latitude, or with decreasing temperatures, although the exact taxonomic scope of this generalization is debated (Meiri and Thomas 2007). Other similar hypotheses exist relating mass evolution to factors such as humidity (Hamilton 1961), productivity (Rosenzweig 1968), and climatic stability (Boyce 1978). There is some evidence that within some species of bats individuals follow Bergmann's rule (individuals growing larger at high latitudes and in low temperatures) (Meiri and Dayan 2003), but this has never been examined explicitly interspecifically, as the existing studies in mammals either omit bats altogether or do not separate them from all other mammals (Rodriguez et al. 2006; Rodriguez et al. 2008).

Examining the patterns within our databases, there is some evidence to support these hypotheses; for example, within all mammals there is evidence that there is an increase in median body size toward higher latitudes on some continents (e.g., in the Americas), or along altitudes (e.g., within the Himalayas)—see figure 4.3*a*. However, there is no clear pattern within bats (fig. 4.3*b*).

We test Bergmann's rule explicitly in bats by correlating ln (body mass) with the absolute value of the latitude of the center of species' distribution. As these variables would be expected to be nonlinearly related, we started with a model containing cubic, quadratic, and linear latitude variables. We used Akaike Information Criterion (AIC) values to select the best minimum adequate model, which was found using stepwise model selection. Additionally, the ln-transformed body mass values were further ln transformed to obtain a normal distribution. Our final model (a combination of the cubic and the quadratic terms of latitude) suggests that latitude and ln (body mass) have a complex negative relationship in bats, exactly opposite to that predicted by Bergmann's rule (fig. 4.4).

However, this result may be biased by the phylogenetic signal in body mass. For example, the family Pteropodidae (flying foxes) are characterized by their large size and all have a tropical distribution, and this may

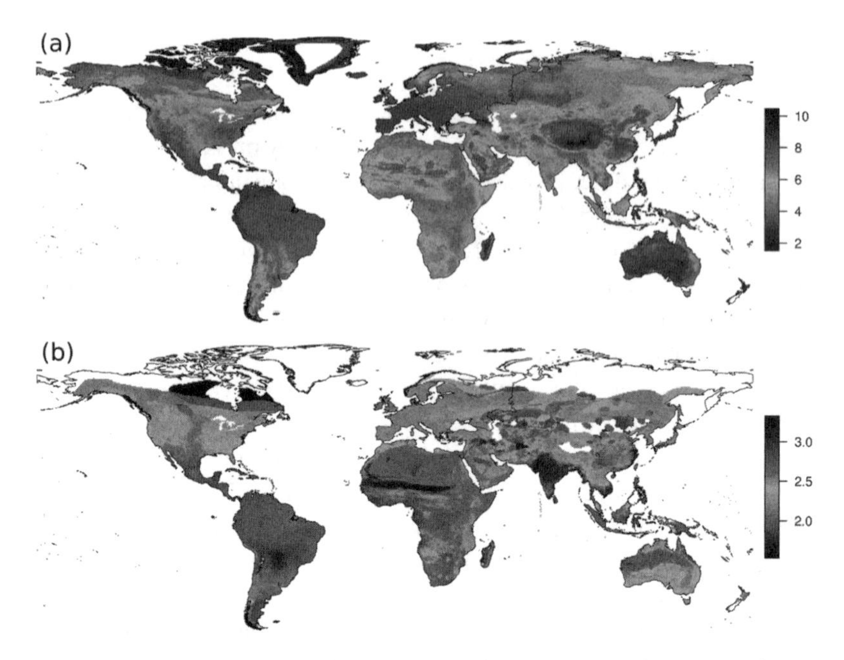


FIGURE 4.3. Maps of median ln (body mass) for (*a*) all mammals (3,841 species), and (*b*) bats (907 species), gridded from species range maps onto 0.5 degree grid cells calculating the median body mass value for all species within that cell.

bias any trend seen. Treating species values as independent will overestimate the degrees of freedom in these analyses, and therefore the degree of phylogenetic signal needs to be taken into account (Harvey and Pagel 1991). Additionally, closely related species tend to be neighbors, so neighboring species will have similar traits, which also make them spatially nonindependent and will bias the model (Harvey and Pagel 1991). Therefore, we repeated the test, controlling for the phylogenetic and spatial bias.

Traditional methods of removing phylogenetic signal from analyses such as CAIC (Purvis and Rambaut 1995) do not control for the amount of variance explained by spatial proximity. Therefore, we used a combination of two recent methods: spatial eigenvector filtering (SEVF) (Diniz-Filho and Bini 2005) to control for spatial biases and phylogenetic eigenvector regression (PVR) (Diniz-Filho et al. 1998; Diniz-Filho 2001) to control for phylogenetic biases. We derived a set of orthogonal vectors from symmetrical pairwise phylogenetic and spatial distance matrices using eigenvector analysis. We calculated the phylogenetic distance

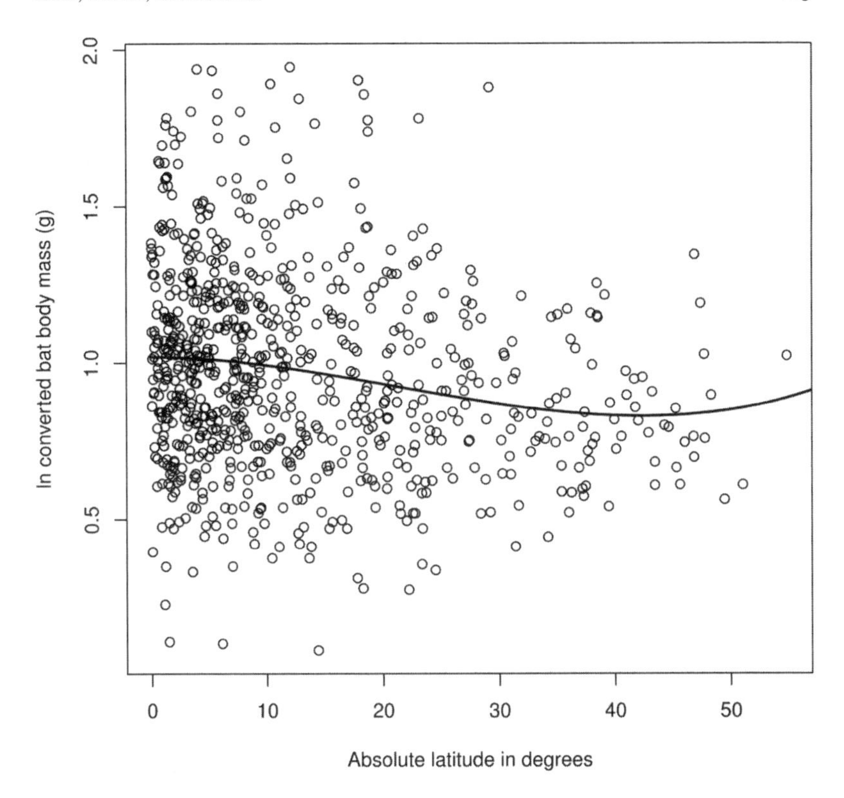


FIGURE 4.4. Relationship between absolute latitude and ln (body mass) in bats. ln (ln (body mass)) = 4.952e − 06 (±2.516e − 06) * |Latitude|$^3$ − 3.155e − 04 (±1.069e − 04) * |Latitude|$^2$ + 1.02. All the terms are significant at $p < 0.05$.

TABLE 4.2 **Tests of Bergmann's Rule in Bats**

| | ln (ln (Body Mass)) vs. Latitude | | |
|---|---|---|---|
| Analysis Type | $R^2$ (%) | $F_{2,773}$ | $p$ |
| Uncorrected data | 3.3 | 25.38 | <0.0001 |
| SEVF only | 1.5 | 6.03 | 0.003 |
| PVR only | 0.3 | 1.1 | 0.33 |
| SEVF and PVR | 0.08 | 0.31 | 0.73 |

matrix using the mammalian supertree (Bininda-Emonds et al. 2007) and generated the spatial distance matrix using the centroids of species distribution ranges (Grenyer et al. 2006). We used all significant phylogenetic and spatial eigenvectors to account for the nonindependence in the species-level data (Desdevises et al. 2003). The efficiency in removing bias was assessed with Moran's *I* test on the residuals of the models. We compared the amount of variance explained in ln (ln (body mass)) as a function of absolute latitude in four analyses, using (1) uncorrected species-level data, (2) data corrected for spatial proximity only (SEVF), (3) data corrected for phylogenetic bias only (PVR), and (4) data corrected for spatial and phylogenetic bias (SEVF and PVR) (table 4.2). The analysis of the explained variation in body mass as a function of latitude suggests that the observed latitudinal gradient in body mass is almost entirely due to phylogenetic bias.

Latitude can be thought of as a crude proxy for environmental conditions, and perhaps using more accurate data may be a more appropriate test of Bergmann's rule. We tested the influence of several additional environmental variables on the variance in body mass: temperature, precipitation, and actual and potential evapotranspiration. We used a 0.5 degree gridded source, *Global GIS* (2003), to extract the relevant environmental data and calculated for each species the mean values that occur within their distributional ranges. The means were obtained by weighting each of these gridded values by the actual area grid cell size (using a Mollweide equal area projection).

We used a stepwise model selection procedure to select out orthogonal environmental variables (using AIC to optimize the number of variables in a minimum adequate model). Of the environmental variables only precipitation and actual evapotranspiration remained in the model (positively correlated) and these explained a significant 3.8% of the variance in body mass ($F_{2,773} = 15.26$; $p < 0.0001$). However, they do not explain any of the remaining residual variation after removing phylogenetic and spatial bias ($r^2 = 0.02$, $F_{2,773} = 0.06$, $p = 0.94$). Similarly to the pattern found in latitude, 1.5% of the variance in residual body mass variation is within the purely phylogenetically explained proportion, and the remaining 2.3% has to be attributed to the overlapping effects of phylogeny and space. To conclude, our analyses demonstrate for the first time no evidence for Bergmann's rule in bats.

We use the eigenvector approach to compare the relative strength of the phylogenetic and spatial component in explaining the variance in bat body mass and compare it to selected mammalian clades (fig. 4.5). Between 39.1% and 53.6 % of the variance in body mass can be explained by phylogeny alone within the four selected clades, compared to between 1.4% and 4% explained by spatial proximity alone. The model containing all the significant phylogenetic and spatial eigenvectors explained

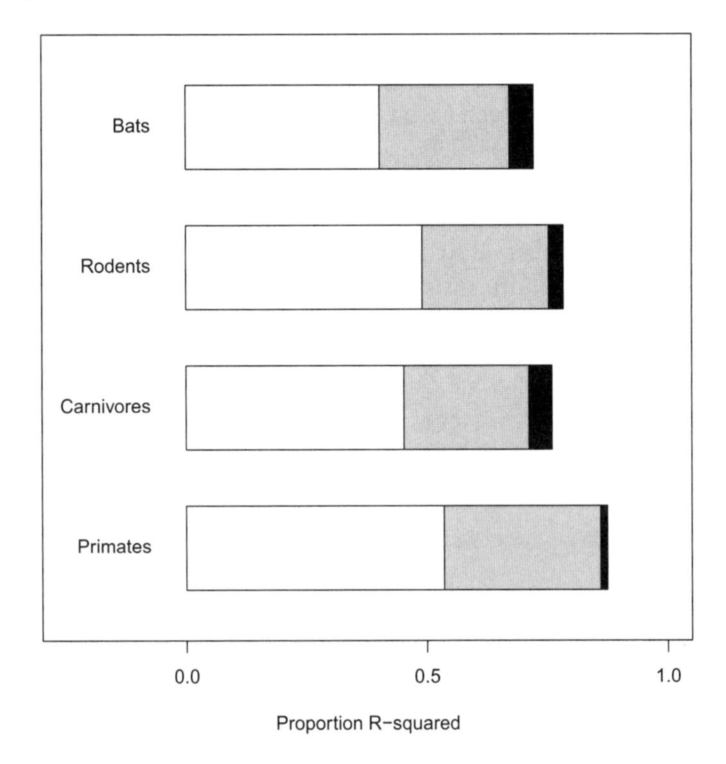


FIGURE 4.5. Decomposition analysis of explained deviance in body mass in mammals. White depicts the pure phylogenetic component (p-component), black is the purely spatial or specific component (s-component), and gray is the overlap between phylogenetic and spatial components.

75.18% of the variance in bat body mass ($F_{47,728} = 46.91$, $p < 0.0001$), with the phylogenetic processes having a more significant effect on the evolution of mass than spatial proximity. Thus the evolution of body mass in bats, despite their ability for powered flight and large dispersal distances, is mainly phylogenetically constrained and shows lower levels of spatial differences. After controlling for phylogeny and spatial proximity, 24.4% of the variance remains in bat body mass. This suggests other adaptive processes have influenced bat body mass evolution, but what these are remains a matter of further study.

Although there seems little evidence for Bergmann's rule in bats according to our analyses, it has been suggested that the results of such studies may depend on differences in how variables are analytically measured within an area (Meiri and Thomas 2007). For example, results can differ depending on whether measures are based on means or medians within assemblages within grid cells or on single values per species, as can be seen by comparing results of Ollalla-Tarraga and Rodriguez (2007) with those of Adams and Church (2008). Species richness (if not appropriately controlled for) may also influence the mean and median sizes within assemblages, because at larger spatial scales and at low latitudes, size frequency distributions tend to be much more modal than the relatively log-uniform distributions found at local scales and high latitudes (Brown and Nicoletto 1991; Cardillo 2002). Species richness was recently found to be a much stronger predictor of median body sizes than were a suite of environmental variables within grid cells in a global analysis of bird body sizes (Olson et al. 2009). However, despite these caveats, there seems to be little signal of a latitudinal or any other environmentally induced mass cline in bats.

### *Island Rule*

Another factor that is widely cited in the literature to determine body mass is island living (Van Valen 1973; Lomolino 2005). Islands are thought to be arenas where body size evolution is both rapid and extreme (Berry 1964; Hooijer 1967; Sondaar 1977; Lister 1989; Millien 2006). Foster (1964) surveyed patterns in mass evolution in several mammalian clades, arguing that some (e.g., Artiodactyls) have a tendency to dwarf on islands, whereas rodents usually increase in size on islands relative to their mainland relatives. Van Valen (1973) interpreted this observation as a general pattern of large mammals dwarfing on islands whereas small mammals evolve larger sizes and referred to it as an ecological "rule." There is some evidence to suggest a tendency toward gigantism in birds (Olson et al. 2009) but very little to support this pattern in bats. This pattern was first examined in 1967 by Krzanowski, who concluded that bats usually dwarf on islands. Lomolino (2005) used Krzanowski's (1967) data to support the notion that the island rule is prevalent within bats: small bats grow larger on islands, whereas larger bats grow smaller. However, using a more comprehensive dataset, Meiri et al. (2008) found that bats as a clade, and subclades within bats, did not tend toward either insular gigantism or insular dwarfism. Moreover, there was no relationship between body sizes of mainland bats and the direction and degree of size evolution of their insular conspecifics, as predicted by the island rule, either in a conventional analysis or when using phylogenetically independent contrasts.

## Conclusions

Bat body mass extends from under 2 g to over 1 kg and shows a typical right-hand skew on a logarithmic scale. In common with other mammalian clades, there is a strong phylogenetic signal in bat body masses where a species' body mass is influenced by the mass of its closest ancestor. Bat body mass has not evolved randomly, and there is evidence of a constraint possibly through the limitations of flight. Evidence suggests that body mass has evolved to be either smaller or larger in different lineages, with the ancestral mass of modern bats estimated to be around 19 g. There is little evidence that body mass has influenced the diversification rate of bats. Mass in bats is less tightly constrained to other biological traits than it is in other mammals, and their life history allometries do not fit the hypothesized optimal models of life history evolution. There is no evidence for either Bergmann's rule or the island rule operating between bat species. Accounting for the influence of both phylogeny and spatial proximity, 24.4% of the variance in bat body mass remains unexplained by any current hypotheses. It is an exciting time for macroevolutionary and macroecology studies of bats, with the growth and availability of global environmental and biodiversity datasets, robust phylogenetic frameworks and analytical methodologies.

## Acknowledgments

We thank Stuart Parsons for comments on earlier versions of the manuscript and acknowledge the Zoological Society of London, the Leverhulme Trust, National Swiss Foundation, and the Max Planck Institute for Ornithology for financial support.

## References

Adams, D. C., and J. O. Church. 2008. "Amphibians do not follow Bergmann's rule." *Evolution* 62:413–420.

Barclay, R. M. R., and R. M. Brigham. 1991. "Prey detection, dietary niche breadth, and body size in bats: Why are aerial insectivorous bats so small?" *American Naturalist* 137:693–703.

Bergmann, K. 1847. "Ueber die verhaltnisse der warmeokonomie der thiere zu ihrer grosse." *Gottinger studien* 3:595–708.

Berry, R. J. 1964. "The evolution of an island population of the house mouse." *Evolution* 18:468–483.

Bininda-Emonds, O. R. P., M. Cardillo, K. E. Jones, R. D. E. MacPhee, R. M. D. Beck, R. Grenyer, S. A. Price, R. Vos, Gittleman J. L., and A. Purvis. 2007. "The delayed rise of present-day mammals." *Nature* 446:507–512.

Blomberg, S. P., T. Garland, and A. R. Ives. 2003. "Testing for phylogenetic signal in comparative data: Behavioral traits are more labile." *Evolution* 57:717–745.

Boyce, M. S. 1978. "Climatic variability and body size variation in the muskrats (*Ondatra zibethicus*) of North America." *Oecologia* 36:1–20.

Brown, J. H., and P. F. Nicoletto. 1991. "Spatial scaling of species composition: Body masses of North American land mammals." *American Naturalist* 138:1478–1512.

Calder, W. A. 1996. *Size, function and life history.* 2nd ed. Cambridge, MA: Harvard University Press.

Cardillo, M. 2002. "Body size and latitudinal gradients in regional diversity of New World birds." *Global Ecology and Biogeography* 11:59–65.

Cardillo, M., C. D. L. Orme, and I. P. F. Owens. 2005. "Testing for latitudinal bias in diversification rates: An example using New World birds." *Ecology* 86:2278–2287.

Charnov, E. L. 1991. "Evolution of life-history variation among female mammals." *Proceedings of the National Academy of Sciences of the United States of America* 88:1134–1137.

——. 1993. *Life-history invariants: Some explorations of symmetry in evolutionary ecology.* Oxford: Oxford University Press.

Clauset, A., and D. H. Erwin. 2008. "The evolution and distibution of species body size." *Science* 321:399–401.

Clauset, A., D. J. Scbwab, and S. Redner. 2009. "How many species have mass M?" *American Naturalist* 173:256–263.

Cooper, N., and A. Purvis. 2010. "Body size evolution in mammals: Complexity in tempo and mode." *American Naturalist* 175 (6): 727–38.

Desdevises, Y., P. Legendre, L. Azouzi, and S. Morand. 2003. "Quantifying phylogenetically structured environmental variation." *Evolution* 57 (11): 2647–2652.

Dial, K. P., and J. M. Marzluff. 1988. "Are the smallest organisms the most diverse?" *Ecology* 69:1620–1624.

Diniz-Filho, J. A. F. 2001. "Phylogenetic autocorrelation under distinct evolutionary processes." *Evolution* 55 (6): 1104–1109.
Diniz-Filho, J. A. F, and L. M. Bini. 2005. "Modelling geographical patterns in species richness using eigenvector-based spatial filters." *Global Ecology and Biogeography* 14:177–185.
Diniz-Filho, J. A. F., C. E. R. De Sant'ana, and L. M. Bini. 1998. "An eigenvector method for estimating phylogenetic inertia." *Evolution* 52 (5): 1247–1262.
Dobson, G. E. 1875. "Conspectus of the suborders, families and genera of Chiroptera arranged according to their natural affinities." *Annuals of the Magazine of Natural History*, 4th ser., 16:345–357.
Felsenstein, J. 1973. "Maximum-likelihood estimation of evolutionary trees from continuous characters." *American Journal of Human Genetics* 25:471–492.
Foster, J. B. 1964. "Evolution of mammals on islands." *Nature* 202:234–235.
Gillman, M. P. 2007. "Evolutionary dynamics of vertebrate body mass range." *Evolution* 61:685–693.
Gittleman, J. L., and M. Kot. 1990. "Adaptation: Statistics and a null model for estimating phylogenetic effects." *Systematic Zoology* 39:227–241.
*Global GIS: A digital atlas of planet earth 6.4.2.* Flagstaff, AZ: U.S. Department of the Interior, U.S. Geological Survey.
Grenyer, R., C. D. L. Orme, S. F. Jackson, G. H. Thomas, R. G. Davies, T. J. Davies, K. E. Jones, V. A. Olson, R. S. Ridgely, P. C. Rasmussen, T.-S. Ding, P. M. Bennett, T. M. Blackburn, K. J. Gaston, J. L. Gittleman, and I. P. F. Owens. 2006. "The global distribution and conservation of rare and threatened vertebrates." *Nature* 444:93–96.
Gunnell, G., and N. B. Simmons. 2005. "Fossil evidence and the origin of bats." *Journal of Mammalian Evolution* 12:209–246.
Hamilton, T. H. 1961. "The adaptive significance of intraspecific trends of variation in wing length and body size among bird species." *Evolution* 15:180–195.
Hansen, T. F. 1997. "Stabilizing selection and the comparative analysis of adaptation." *Evolution* 51:1341–1351.
Harvey, P. H., and M. D. Pagel. 1991. *The comparative method in evolutionary biology.* Oxford: Oxford University Press.
Hooijer, D. A. 1967. "Indo-Australian insular elephants." *Genetica* 38:143–162.
Hutcheon, J. M. , and J. A. W. Kirsch. 2006. "A moveable face: Deconstructing the Microchiroptera and a new classification of extant bats." *Acta Chiropterologica* 8:1–10.
Isaac, N. J. B., K. E. Jones, J. L. Gittleman, and A. Purvis. 2005. "Correlates of species richness in mammals: Body size, life history, and ecology." *American Naturalist* 165:600–607.
Jones, K. E., J. Bielby, M. Cardillo, S. A. Fritz, J. O'Dell, C. D. L. Orme, K. Safi, W. Sechrest, E. H. Boakes, C. Carbone, C. Connolly, M.1 J. Cutts, J. K. Fos-

ter, R. Grenyer, M. Habib, C. A. Plaster, S. A. Price, E. A. Rigby, J. Rist, A. Teacher, O. R. P. Bininda-Emonds, J. L. Gittleman, G. M. Mace, and A. Purvis. 2009. "PanTHERIA: A species-level database of life-history, ecology and geography of extant and recently extinct mammals." *Ecology* 90:2648.

Jones, K. E, and A. MacLarnon. 2001. "Bat life-histories: Testing models of mammalian life history evolution." *Evolutionary Ecology Research* 3:465–476.

———. 2004. "Affording larger brains: Testing hypotheses of mammalian brain size evolution on bats." *American Naturalist* 164:20–31.

Jones, K. E., and A. Purvis. 1997. "An optimum body size for mammals? Comparative evidence from bats." *Functional Ecology* 11:751–756.

Koopman, K. F. 1984. "Bats." In *Orders and families of recent mammals of the world.*, edited by S. Anderson and J. K. Jones, 145–186. New York: John Wiley and Sons.

Kozlowski, J., and J. Weiner. 1997. "Interspecific allometries are by-products of body size optimization." *American Naturalist* 149:352–380.

Krzanowski, A. 1967. "The magnitude of islands and the size of bats (Chiroptera)." *Acta Zoologica Cracoviensia* 12:281–346.

Lindenfors, P., J. L. Gittleman, and K. E. Jones. 2007. "Sexual size dimorphism in mammals." In *Sex, Size and Gender Roles*, edited by D. J. Fairburn, W. U. Blanckenhorn, and T. Szekely, 16–26. Oxford: Oxford University Press.

Lister, A. M. 1989. "Rapid dwarfing of red deer on Jersey in the last interglacial." *Nature* 342:539–542.

Lomolino, M. V. 2005. "Body size evolution in insular vertebrates: Generality of the island rule." *Journal of Biogeography* 32:1683–1699.

Maltby, A., K. E. Jones, and G. Jones. 2010. "Understanding the origination and diversification of bat echolocation calls." In *Handbook of mammalian vocalization. An integrative neuroscience approach*, edited by S. M. Brudzynski, 37–47. London: Elsevier.

Mattila, T. M., and F. Bokma. 2008. "Extant mammal body masses suggest punctuated equilibrium." *Proceedings of the Royal Society B: Biological Sciences* 275:141–148.

Mayr, E. 1956. "Geographical character gradients and climatic adaptation." *Evolution* 10:105–108.

Meiri, S., N. Cooper, and A. Purvis. 2008. "The island rule: Made to be broken?" *Proceedings of the Royal Society B: Biological Sciences* 275:141–148.

Meiri, S., and T. Dayan. 2003. "On the validity of Bergmann's rule." *Journal of Biogeography* 30:331–351.

Meiri, S., and G. H. Thomas. 2007. "The geography of body size: Challenges of the interspecific approach." *Global Ecology and Biogeography* 16:689–693.

Millien, V. 2006. "Morphological evolution is accelerated among island mammals." *PLoS Biology* 4:1863–1868.

Murphy, W. J., E. Eizirik, S. J. O'Brien, O. Madsen, M. Scally, C. J. Douady,

E. Teeling, O. A. Ryder, M. J. Stanhope, W. W. de Jong, and M. S. Springer. 2001. "Resolution of the early placental mammal radiation using Bayesian phylogenetics." *Science* 294:2348–2351.

Neuweiler, G. 2000. *The biology of bats*. New York, Oxford: Oxford University Press.

Norberg, U M. 1987. "Wing form and flight mode in bats." In *Recent advances in the study of bats*, edited by M Brock Fenton, P Racey, and J. M. V. Rayner, 43–56. Cambridge: Cambridge University Press.

———. 1994. "Wing design, flight performance, and habitat use in bats." In *Ecological morphology*, edited by P. C. Wainwright and S. M. Reilly, 205–239. Chicago: University of Chicago Press.

Norberg, U. M., and J. M. V. Rayner. 1987. "Ecological morphology and flight in bats (Mammalia, Chiroptera): Wing adaptations, flight performance, foraging strategy and echolocation." *Philosophical Transactions of the Royal Society B: Biological Sciences* 316 (1179): 337–419.

Olden, J. D., Z. S. Hogan, and M. J. Vander Zanden. 2007. "Small fish, big fish, red fish, blue fish: size-biased extinction risk of the world's freshwater and marine fishes." *Global Ecology and Biogeography* 16:694–701.

Ollalla-Tarraga, M., and M. A. Rodriguez. 2007. "Energy and interspecific body size patterns of amphibian faunas in Europe and North America: Anurans follow Bergmann's rule, urodeles its converse." *Global Ecology and Biogeography* 16:606–617.

Olson, V. A., R. G. Davies, C. D. L. Orme, G. H. Thomas, S. Meiri, T. M. Blackburn, K. J. Gaston, I. P. F. Owens, and P. M. Bennett. 2009. "Global biogeography and ecology of body size in birds." *Ecology Letters* 12:249–259.

Orme, C. D. L., N. J. B. Isaac, and A. Purvis. 2002. "Are most species small? Not within species-level phylogenies." *Proceedings of the Royal Society, B: Biological Sciences* 269:1279–1287.

Pagel, M. D. 1999. "Inferring the historical pattern of biological evolution." *Nature* 401:877–884.

Pettigrew, J. D. 1986. "Flying primates: Megabats have the advanced pathway from eye to midbrain." *Science* 231:1304–1306.

———. 2008. "Primate-like retinotectal decussation in an echolocating megabat, *Rousettus aegyptiacus*." *Neuroscience* 153:226–231.

Purvis, A., and P. H. Harvey. 1995. "Mammal life-history evolution: A comparative test of Charnov's model." *Journal of Zoology* 237:259–283.

Purvis, A., C. D. L. Orme, and K. Dolphin. 2003. "Why are most species small-bodied? A phylogenetic view." In *Macroecology: Concepts and consequences*, edited by T. M. Blackburn and K. J. Gaston. Oxford: Blackwell Scientific.

Purvis, A., and A. Rambaut. 1995. "Comparative analysis by independent contrasts (CAIC): An Apple Macintosh application for analysing comparative data." *Computer Applications for BioSciences* 11:247–251.

Ralls, K. 1976. "Mammals in which females are larger than males." *Quarterly Review of Biology* 51 (2): 245–276.

Rayner, J. M. V. 1996. "Biomechanical constraints on size in flying vertebrates." *Symposium of the Zoological Society of London* 69:83–109.

R Development Core Team. 2008. *R: A language and environment for statistical computing.* Vienna: R Foundation for Statistical Computing.

Rodriguez, M. A., I. L. Lopez-Sanudo, and B. A. Hawkins. 2006. "The geographic distribution of mammal body size in Europe." *Global Ecology and Biogeography* 15:173–181.

Rodriguez, M. A., M. A. Olalla-Tarraga, and B. A. Hawkins. 2008. "Bergmann's rule and the geography of mammal body size in the Western Hemisphere." *Global Ecology and Biogeography* 15:173–181.

Rosenzweig, M. L. 1968. "Net primary productivity of terrestrial communities: Prediction from climatological data." *American Naturalist* 102:67–74.

Safi, K., and D. K. N. Dechmann. 2005. "Adaptation of brain regions to habitat complexity: A comparative analysis in bats (Chiroptera)." *Proceedings of the Royal Society B: Biological Sciences* 272:179–186.

Simmons, N. B., and T. M. Conway. 2003. "Evolution of ecological diversity in bats." In *Bat ecology*, edited by T. H. Kunz and M. B. Fenton, 493–535. Chicago: University of Chicago Press.

Smith, F. A., J. H. Brown, J. P. Haskell, S. K. Lyons, J. Alroy, E. L. Charnov, T. Dayan, B. J. Enquist, S. K. M. Ernest, E. A. Hadly, K. E. Jones, D. M. Kaufman, P. A. Marquet, B. A. Maurer, K. J. Niklas, W. P. Porter, B. Tiffney, and M. R. Willig. 2004. "Similarity of mammalian body size across the taxonomic hierarchy and across space and time." *American Naturalist* 163:672–691.

Sondaar, P. Y. 1977. "Insularity and its effects on mammal evolution." In *Major patterns of vertebrate evolution*, edited by M. K. Hecht, P. C. Goody, and B. M. Hecht, 671–707. New York: Plenum Press.

Springer, M. S., E. C. Teeling, O. Madsen, M. J. Stanhope, and W. W. de Jong. 2001. "Integrated fossil and molecular data reconstruct bat echolocation." *Proceedings of the National Academy of Sciences of the United States of America* 98:6241–6246.

Stanley, S. M. 1973. "An explanation for Cope's rule." *Evolution* 27:1–26.

Stearns, S. C. 1992. *The Evolution of Life Histories.* Oxford: Oxford University Press.

Teeling, E. C., M. Scally, D. J. Kao, M. L. Romagnoli, M. S. Springer, and M. J. Stanhope. 2000. "Molecular evidence regarding the origin of echolocation and flight in bats." *Nature* 403:188–192.

Teeling, E. C., M. S. Springer, O. Madsen, P. Bates, S. J. O'Brien, and W. J. Murphy. 2005. "A molecular phylogeny for bats illuminates biogeography and the fossil record." *Science* 307:580–584.

Van Valen, L. M. 1973. "Body size and numbers of plants and animals." *Evolution* 27:27–35.

Webster, A., and A. Purvis. 2001. "Testing the accuracy of methods for reconstructing ancestral states of continuous characters." *Proceedings of the Royal Society B: Biological Sciences* 269:143–149.

———. 2002. "Ancestral states and evolutionary rates of continuous characters." In *Morphology, shape and phylogenetics*, edited by N. MacLeod and P. Forey. London: Taylor and Francis.

Williams, D. F., and J. S. Findley. 1979. "Sexual size dimorphism in Vespertilionid bats." *American Midland Naturalist* 102 (1): 113–127.

Willig, M. R., B. D. Patterson, and R. D. Stevens. 2003. "Patterns of range size, richness, and body size in the Chiroptera." In *Bat ecology*, edited by T. H. Kunz and B. M. Fenton. Chicago: University of Chicago Press.

Wilson, D. E., and D. M. Reeder. 1993. *Mammal species of the world: A taxonomic and geographic reference*. 2nd ed. Washington, DC: Smithsonian Institution Press.

CHAPTER FIVE

# Macroecological Patterns of Body Size in Mammals across Time and Space

S. Kathleen Lyons and Felisa A. Smith

Macroecology is a way of thinking, exploring, and asking questions about complex ecological phenomena. By examining large-scale patterns across temporal, spatial, or taxonomic scales, consistent, common patterns emerge that suggest common underlying causes and processes. The field is expanding rapidly, and macroecological approaches to biology have led to new insights into the operation of general mechanistic processes that govern the structure and dynamics of individuals, populations, and complex ecological systems (Smith et al. 2008).

Here we are using a macroecological approach to examine patterns in mammalian body size distributions at multiple spatial and temporal scales. Body size is arguably one of the most fundamental attributes of an organism and has long been of interest to scientists. Many physiological and ecological traits scale with body size (Brown 1995). As a result, information about the body size of an organism can provide considerable information about its ecology and life history. There is enormous variation in the body size of extant organisms, encompassing approximately twenty-four orders of magnitude (Peters 1983; Calder 1984; Niklas 1994), and within mammals body size spans approximately ten orders of magnitude (Brown and Nicoletto 1991; Brown 1995; Alroy 1998; Marquet and Cofre 1999; Bakker and Kelt 2000; Lyons et al. 2004; Smith et al. 2004).

Much effort has focused on the consequences of being a certain size and whether taxa exhibit an optimal size (e.g., Brown and Nicoletto 1991;

Brown et al. 1993; Alroy 1998; Smith et al. 2004), and as a result, mammalian body size distributions have been well studied and found to be remarkably consistent across continents (Smith et al. 2004) and across recent time (Lyons et al. 2004). Despite these comprehensive studies, there are still holes in our understanding of the macroecology of body size. For example, the largest continent (i.e., Eurasia) has not been analyzed, and the shapes of continental body size distributions have not been evaluated across the entire Cenozoic. Community body size distributions have been studied on three of the major continents (i.e., North America [Brown and Nicoletto 1991], South America [Marquet and Cofre 1999; Bakker and Kelt 2000], and Africa [Kelt and Meyer 2009]) but have not been evaluated on Eurasia and Australia or across time. Moreover, other macroecological patterns of body size, such as the relationship between body size and range size, have not been evaluated using fossil data.

What follows is a synthetic analysis of what is known about macroecological patterns of body size in mammals. We review known patterns and extend them either spatially to previously unexamined continents or temporally into the fossil record. The majority of the chapter will focus on patterns of body size distributions at multiple spatial and temporal scales. In addition to reviewing the pertinent literature, we include new data and analyses derived from our own investigations. Building on our earlier work (Lyons et al. 2004; Smith et al. 2004), we extend our analyses to include Eurasia and a number of large islands. Moreover, we include an analysis of the shape of continental body size distribution for North American mammals across the last 65 million years. In addition, we will also include an analysis of the relationship between geographic range size and body size in mammals, and we will examine how that relationship fluctuates over time. Finally, we compare the shapes of community-level body size distributions on multiple continents and in the late Pleistocene of North America.

Taken together, the data and analyses presented herein attempt to address several questions about the macroecology of body size in mammals. First, how similar are overall body size distributions across individual continents and large islands? Second, how similar are overall body size distributions across time in North America? Are present-day patterns warped? Third, is the relationship between body size and range size consistent across time and space? Finally, are the shapes of body size distributions at the community level consistent across time and space?

Although many of these questions have been addressed previously, to our knowledge, this is the most comprehensive analysis, both spatially and temporally, of the macroecological patterns of body size to date.

## Data and Analyses

The body mass data used for this review come from MOM v. 3.6, an updated version of Smith et al. (2003) that is available from either author upon request. The taxonomy used is Wilson and Reeder (1993). This dataset includes body mass estimates for the majority of the mammals of the world and the large islands. It includes estimates for extant species, species that have gone extinct in historical times, and species that went extinct during the megafaunal extinction in the late Pleistocene. Methods for data collection and handling can be found in Smith et al. (2003).

Here, we use generic averages for species without body mass estimates in order to fully characterize body size distributions. Our final dataset had estimates for 833 species from Africa, 813 species from Eurasia, 600 species from North America, 777 species from South America, 256 species from Australia, 40 species from Tasmania, 116 from Madagascar, and 150 species from New Guinea. Volant, introduced, and marine mammals were excluded from all analyses.

For each continent and large island, frequency distributions of body sizes were created and descriptive statistics were calculated for both modern (table 5.1) and late Pleistocene faunas (table 5.2). The effect of the missing values was evaluated by comparing the shapes of the body size distributions on each continent and large island with and without generic averages. Both Kolmorogorov-Smirnov two-sample tests and Mann-Whitney U tests were used. For the majority of the distributions, there was no difference between the shapes of the distributions with and without generic estimates for missing values (table 5.3). However, the shapes of the distributions for Eurasia and Africa were significantly different with and without the estimates for missing values. In each case, the missing species were primarily rare rodents and shrews and were concentrated in the lower end of the body size distribution. Because the majority of mammal genera encompass less than one order of magnitude of variation in log mass, including the estimates for missing

TABLE 5.1 **Descriptive Statistics for Body Size Distributions of Extant Mammals for Each Continent and Island Separately and for All Continental Mammals Together**

| Land Mass | N | SD | Mean | Median | Mode | Range | Min | Max | Skewness | Kurtosis |
|---|---|---|---|---|---|---|---|---|---|---|
| All continents | 3,057 | 1.160 | 2.378 | 2.059 | 1.130 | 6.353 | 0.243 | 6.596 | 0.876 | 0.153 |
| Africa | 820 | 1.314 | 2.354 | 1.875 | 1.130 | 6.353 | 0.243 | 6.596 | 0.893 | −0.187 |
| Eurasia | 802 | 1.277 | 2.523 | 2.176 | 1.547 | 6.435 | 0.371 | 6.435 | 0.742 | −0.304 |
| North America | 522 | 1.035 | 2.176 | 1.926 | 1.602 | 5.383 | 0.380 | 5.763 | 0.971 | 0.843 |
| South America | 701 | 0.895 | 2.380 | 2.212 | 2.445 | 5.009 | 0.672 | 5.681 | 0.930 | 0.493 |
| Australia | 211 | 1.032 | 2.402 | 2.121 | — | 4.037 | 0.628 | 4.665 | 0.347 | −1.095 |
| Tasmania | 30 | 1.022 | 2.770 | 2.990 | — | 3.800 | 0.850 | 4.650 | −0.048 | −0.998 |
| Madagascar | 102 | 1.060 | 2.398 | 2.235 | — | 5.445 | 0.530 | 5.975 | 0.604 | 0.274 |
| New Guinea | 140 | 0.831 | 2.487 | 2.385 | — | 3.170 | 1.010 | 4.180 | 0.265 | −0.972 |

TABLE 5.2 **Descriptive Statistics for the Body Size Distributions of Late Pleistocene Mammals for Each Continent and Island Separately and for All Continental Mammals**

| Land Mass | Land Area ($km^2$) | N | SD | Mean | Median | Mode | Range | Min | Max | Skewness | Kurtosis |
|---|---|---|---|---|---|---|---|---|---|---|---|
| All continents | 134,344,000 | 3,279 | 1.300 | 2.568 | 2.161 | 1.130 | 6.757 | 0.243 | 7.00 | 0.907 | 0.055 |
| Africa | 30,065,000 | 833 | 1.364 | 2.403 | 1.903 | 1.130 | 6.570 | 0.243 | 6.813 | 0.897 | −0.179 |
| Eurasia | 54,517,000 | 813 | 1.325 | 2.568 | 2.204 | 1.547 | 6.369 | 0.371 | 6.740 | 0.780 | −0.214 |
| North America | 24,256,000 | 600 | 1.415 | 2.526 | 2.123 | 1.602 | 6.620 | 0.380 | 7.00 | 1.050 | 0.358 |
| South America | 17,819,000 | 777 | 1.279 | 2.683 | 2.360 | 2.445 | 6.208 | 0.672 | 6.880 | 1.219 | 0.873 |
| Australia | 7,687,000 | 256 | 1.364 | 2.847 | 2.708 | 5.176 | 5.548 | 0.628 | 6.176 | 0.395 | −0.903 |
| Tasmania | 68,332 | 40 | 1.304 | 3.290 | 3.265 | — | 5.030 | 0.850 | 5.880 | 0.030 | −0.875 |
| Madagascar | 581,540 | 116 | 1.228 | 2.644 | 2.545 | 4.00 | 5.445 | 0.530 | 5.975 | 0.563 | −0.118 |
| New Guinea | 786,000 | 150 | 0.967 | 2.624 | 2.540 | — | 4.470 | 1.010 | 5.480 | 0.544 | −0.312 |

*Note*: Area of major continents from www.enchantedlearning.com/geography/continents/Land.shtml. Tasmania, Madagascar, and New Guinea from Wikipedia.com. The land area for "All continents" is the sum of Africa, Eurasia, North America, South America, and Australia.

TABLE 5.3 **Comparisons of Mammalian Body Size Distributions with and without Generic Averages for Missing Values**

| Land Mass | No. of Generic Values | $\chi^2$ | P | Z | P |
|---|---|---|---|---|---|
| North America | 28 | 0.150 | >0.999 | −0.119 | 0.905 |
| South America | 3 | 0.052 | >0.999 | −0.128 | 0.898 |
| Eurasia | 162 | 10.224 | 0.012 | −3.346 | <0.001 |
| Africa | 202 | 21.655 | <0.001 | −4.418 | <0.001 |
| Australia | 3 | 0.018 | >0.999 | −0.023 | 0.982 |
| New Guinea | 23 | 0.809 | >0.999 | −0.599 | 0.549 |
| Tasmania | 0 | — | — | — | — |
| Madagascar | 3 | 0.667 | >0.999 | −0.234 | 0.815 |

*Note*: Analyses were performed using nonparametric Kolmorgorov-Smirnov and Mann-Whitney U tests.

values should not change the shape of the binned curve (Smith and Lyons, unpublished data). Because of that, all analyses were performed on the more complete dataset that included estimates for species with missing data.

For deep-time analyses, dates and body masses of North American mammals come from the Paleobiology Database (www.pbdb.org) and from Alroy (1998, 2000). For each of four time periods (3, 20, 40, and 60 million years ago) 1-million-year time slices around those dates were taken and body size distributions of all species extant during each time period were created.

Latitudinal range sizes of extant North American mammals were obtained from an analysis of the effects of incomplete sampling on estimates of the relationship between geographic range size and body size (Madin and Lyons 2005) originally derived from (Patterson et al. 2004). Log range size was plotted as a function of log body mass, and each order of mammals was given a unique symbol. For late Pleistocene mammals, the geographic range sizes and time periods were obtained from an analysis of mammalian geographic range shifts in response to glaciation (Lyons 2003, 2005). The time periods used here (Holocene: 10,000–500 BP; Glacial: 20,000–10,000 BP; and Pre-Glacial 40,000–20,000 BP) are the same as those used by Lyons (2003, 2005) and encompass the expansion and retreat of the ice sheets during the last glaciation. The raw data are available from the FAUNMAP database (FAUNMAP Working Group 1994). In this case, geographic range sizes were calculated in $km^2$ and not degrees of latitudinal extent. As with extant mammals, log range size was plotted as a function of log body mass for each of the three time periods.

For the analyses of local community body size distributions, data and figures come from an analysis of the shapes of body size distributions over space and time (Lyons 2007). Lists of species in local communities are taken from an unpublished dataset compiled by SKL that contains 328 extant communities taken from a variety of literature and web sources, and 328 late Pleistocene and Holocene communities derived mainly from the FAUNMAP database (FAUNMAP Working Group 1994). Fossil communities are divided into late Pleistocene and Holocene communities to evaluate the effects of the megafaunal extinction on community body size distributions. Mean community age for near-time communities was calculated as the mean of the minimum and maximum age estimates for each locality as reported by FAUNMAP. For each local community, descriptive statistics that characterize the shape of the body size distribution were calculated (i.e., mean, median, skewness, and kurtosis). Because the patterns for mean and median body size were not substantially different, only median body size is reported.

Each moment of the body size distribution provides unique information about the overall shape of the distribution, and comparison of the moments gives information about the similarity or difference in different aspects of that shape. In particular, the skewness and kurtosis are extremely useful in concisely describing the shape of a distribution. The skewness measures the length of the curve relative to its height, whereas the kurtosis describes the height of the curve relative to the standard deviations. Moreover, the values of skewness and kurtosis are easily translatable to shape. For the skewness, positive values indicate right skew, negative values indicate left skew. For kurtosis, positive values indicate a peaked distribution with values around zero indicating a normal distribution. Negative values between 0 and −1 indicate a flat distribution and highly negative values greater than −1.5 indicate a bimodal distribution.

The moments of the distributions for near-time communities were compared to that of modern North American communities using *t*-tests to determine if the shapes of local body size distributions were significantly different because of the addition of the extinct megafauna or in the nonanalogue communities associated with glaciation. In addition, the moments describing modern communities on multiple continents were compared using *t*-tests to determine if there are significant differences among the continents.

## Results and Discussion

The analyses presented herein are the most comprehensive to date of macroecological patterns of body size in mammals across space and time. We found that at the level of continents and large islands, there was a remarkable consistency of body size distributions across both time and space that was disrupted by the end-Pleistocene extinctions. These analyses suggest that modern body size distributions for mammals at large spatial scales have been artificially distorted by the activities of aboriginal invaders. We also find a remarkable consistency in the relationship between body size and range size across time both before and after the extinction. Finally, we find that the shapes of community body size distributions vary across both space and time and suggest that further analyses focus on the role of habitat type rather than continental associations.

### *Continental Patterns in Body Size*

HOW SIMILAR ARE OVERALL BODY SIZE DISTRIBUTIONS ACROSS INDIVIDUAL CONTINENTS AND LARGE ISLANDS? Despite the greater amount of data available for this analysis, our results are consistent with Smith et al. (2004). The overall shapes and ranges of the body size distributions are similar on each of the four main continents, Eurasia, Africa, North America, and South America (fig. 5.1, *white bars*). We find that the statistics that describe the body size distributions of the major continents are remarkably similar (table 5.1). The range of body sizes found on each continent differs only slightly with most of the difference attributable to the lack of elephants in the New World and the lack of insectivores in South America. Moreover, the mean and median body sizes for each continent fall within a narrow range (mean: 2.176–2.523 log units, ~150–333 g; median 1.875–2.212 log units, ~75–163 g). Even the skewness is similar, ranging from 0.742 to 0.930. However, there are noticeable differences. Although the distributions are multimodal on each continent, the position of the modes differs. Eurasia and Africa have very similar shapes, with a species-rich second mode of large-bodied mammals that is lacking in modern North and South America. In addition, the position of the small-bodied mode differs on the different continents, in part, due to the order Insectivora. Africa has a large num-

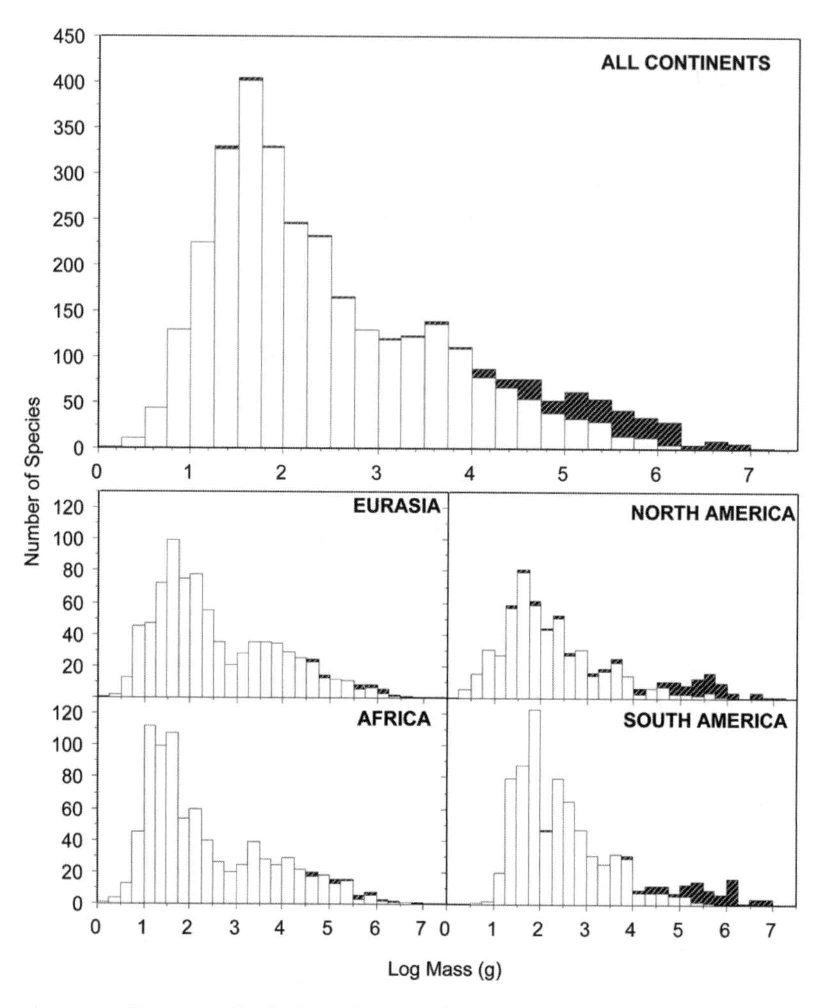


FIGURE 5.1. Frequency distributions of mammalian body size on all continents and four major continents, Eurasia, North America, Africa, and South America. Hatched bars indicate the body size distribution of species that went extinct in the late Pleistocene. Note that the timing of the extinction is not the same on all the continents. The extinction on Eurasia occurred in two pulses at 50,000 and 30,000 years ago (Stuart 1999).

ber of insectivores, causing the mode of small-bodied species to be shifted to the left relative to other continents. In contrast, insectivores do not make it into South America, and as a consequence the mode of small-bodied species is shifted to the right relative to the other continents (see Smith et al. 2004). As a result, there are statistically significant differences between the continents, despite the overall similarity in shape.

The one small continent and the large islands (e.g., Australia, Tasmania, New Guinea, and Madagascar) have body size distributions that are distinct from those of the four main continents (fig 5.2, *white bars*). Rather than having a right-skewed distribution with a mode of small-bodied mammals, large islands have multimodal distributions that are much flatter than continental distributions. Indeed, they strongly resemble the regional distributions of Brown and Nicolleto (1991). Moreover, the degree to which the distribution is flattened appears related to island area and number of species. Tasmania has a much flatter distribution than the other large islands and also contains the fewest species (fig. 5.2, *white bars*, table 5.1) and the smallest land area (table 5.2). As with the continental distributions, the range, mean, and median of the body size distributions of the large islands are relatively similar. Moreover, as with the continents, the differences can be attributed to unique aspects of the islands' history. For example, Madagascar has a much greater body size

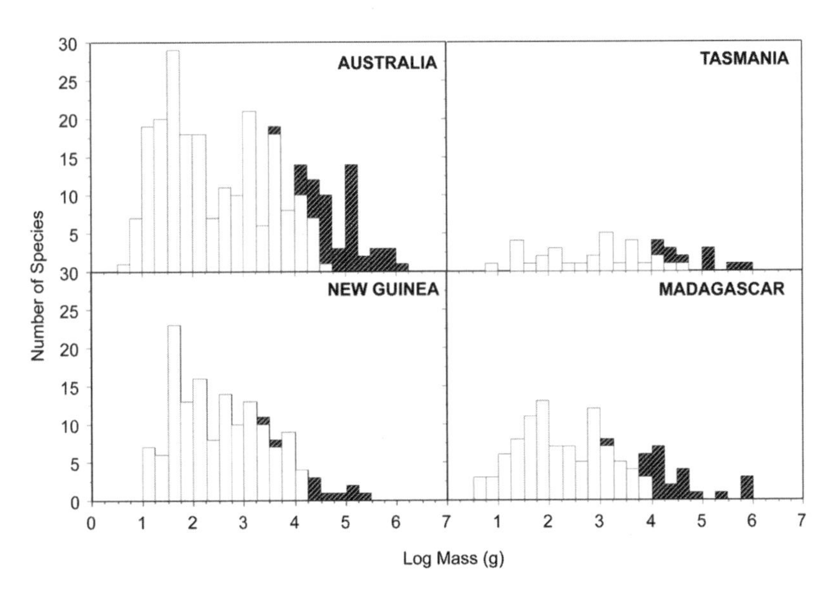


FIGURE 5.2. Frequency distributions of mammalian body size on one small continent, Australia, and three large islands, New Guinea, Tasmania, and Madagascar. Hatched bars indicate the body size distribution of species that went extinct in the late Pleistocene. Note that the timing of the extinction is not the same on all land masses, but each coincides roughly with the arrival of humans (Martin and Steadman 1999).

range because of the inclusion of the endemic pygmy hippo, *Hexaprotodon madagascariensis*. There is some question as to the timing of the extinction of this species. However, evidence suggests that the extinction occurred in historical times (MacPhee and Flemming 1999), thus warranting inclusion in a characterization of the body size distribution of modern mammals of Madagascar.

HOW SIMILAR ARE OVERALL BODY SIZE DISTRIBUTIONS ACROSS TIME IN NORTH AMERICA? On each continent and large island, there was a size-biased extinction during the late Pleistocene that resulted in a truncation of the body size distribution (figs. 5.1 and 5.2, *hatched bars*). North and South America were strongly affected by the extinction event, losing ~70–80 species each, whereas the extinction events were relatively minor in Eurasia and Africa, with each losing ~15 species. Prior to the extinction event, the overall shapes of the continental distributions were even more similar than at present (fig. 5.1). North and South America had a second mode of large-bodied species and a large-bodied fauna that rivaled that of Africa. Similar to Lyons et al. (2004), we find remarkable similarity in the descriptive statistics that characterize the body size distributions on each continent. The mean and median body sizes fall within a narrow range of values (table 5.2). In addition, the differences in the overall range of body masses found on each continent in the present disappear when considering distributions that include the extinct megafauna. Prior to the extinction the body mass distributions on the four main continents encompassed 6.2–6.6 log units of mass. Lyons et al. (2004) found that prior to the extinction event, the shapes of the body size distributions of North and South America were not significantly different from that of Africa, although they were significantly different from each other. Here we find that all of the continents, except North America and Eurasia, are significantly different from one another even with the addition of the extinct megafauna (table 5.4). The difference between our results and that of Lyons et al. (2004) are probably due to the more complete dataset and the larger sample sizes available for this analysis. Moreover, as with the modern distributions, there are unique aspects of each continent's history that will contribute to significant differences between them.

Each of the large islands we examined also had a significantly size-biased extinction that truncated the body size distribution at the large end (fig. 5.2). Interestingly, despite the similarity in the extinction event,

TABLE 5.4 **Comparisons of Mammalian Body Size Distributions on Different Continents prior to the Extinction of the Megafauna**

| | Africa | Eurasia | North America | South America | Australia |
|---|---|---|---|---|---|
| Africa | — | <0.001 | <0.001 | <0.001 | <0.001 |
| | | <0.001 | 0.018 | <0.001 | <0.001 |
| Eurasia | 24.921 | — | 0.133 | <0.001 | <0.001 |
| | −3.507 | | 0.302 | 0.012 | 0.004 |
| North America | 15.670 | 5.429 | — | <0.001 | <0.001 |
| | −2.377 | −1.032 | | <0.001 | <0.001 |
| South America | 78.613 | 22.298 | 19.530 | — | <0.001 |
| | −6.555 | −2.514 | −3.608 | | 0.128 |
| Australia | 29.090 | 16.660 | 24.553 | 23.227 | — |
| | −5.001 | −2.849 | −3.430 | −1.523 | |

*Note*: Analyses were performed using nonparametric Kolmorgorov-Smirnov and Mann-Whitney U tests. Values in the bottom triangle are test statistics, and values in the upper triangle are $p$ values. Upper values are for the Kolmorgorov-Smirnov test, and lower values are for the Mann-Whitney U test. Continent-specific generic averages were substituted for missing body sizes.

the extinction did not occur at the same time on each island (Grainger et al. 1987; Flannery 1995; Strahan 1995; Bonaccorso 1998 ; Garbutt 1999). Indeed, for Madagascar, there is some debate concerning the timing of the extinction of the different species that were affected (Garbutt 1999; MacPhee and Flemming 1999). However, as with the continental distributions, the shapes and the statistics that characterize the body size distributions of mammals on large islands were similar prior to the extinction event (fig. 5.2, table 5.2). In particular, the range of body masses on each island was more similar prior to the extinction event despite the differences in land area (compare table 5.2 to table 5.1). Interestingly, for island distributions, the megafaunal extinction events did not substantially change the shape of the distribution. Because the distributions tend to be flatter and less skewed in general, the removal of large-bodied species simply truncated the distributions rather than removing a second mode, as with the continental distributions.

At the continental level, body size distributions of modern mammals and late Pleistocene mammals are significantly different (Lyons et al. 2004). The lack of megafauna in present-day faunas has had a profound effect on the shape of body size distributions. The question then becomes, Which time period is unique? Are present-day distributions unusual or is there something unique about the late Pleistocene? To answer these questions, we examined the shape of body size distributions for mammals in North America in 1-million-year intervals at four different time periods over the Cenozoic (4, 20, 40, and 60 million years ago).

Because these distributions were generated entirely from fossil data, they are unlikely to be perfect representations of the faunas of the time. First, sampling is a problem, and not all species present at a particular time will be represented in the fossil record. Second, this bias is likely to be stronger for small-bodied species. An analysis of the late Pleistocene record of North America has shown that small-bodied species are less likely to be represented, either because they are less robust and therefore less likely to be fossilized or because they are less likely to be recovered without special collecting techniques (Lyons and Smith 2006a, 2010). However, these biases should be similar throughout the Cenozoic, making comparison among the deep-time faunas possible. Despite these problems, there are similarities between these deep-time faunas and the late Pleistocene that suggest general patterns. First, after the Cretaceous/Paleogene extinction, mammals very quickly filled the full range of body sizes occupied by mammals (fig. 5.3, *bottom panel*; see also Alroy 1998; Smith et al. 2004). Moreover, the mean body masses for the different orders of mammals stayed relatively constant over time and across continents (Smith et al. 2004). Second, the body size distribution of mammals became bimodal by 40 million years ago and stayed that way until the extinction of the megafauna in the late Pleistocene (fig. 5.3; see also Alroy 1998).

The similarity of extinction events on multiple continents and large islands suggests that there should be predictability in the shape of mammalian body size distributions at continental and regional scales. Continental body size distributions should be right skewed and multimodal with two major modes—one for small-bodied species and one for the large-bodied species that are lacking from many modern faunas. Indeed, mammalian body size distributions became bimodal early on in the history of mammals, first occurring around 40 million years ago. Moreover, our results for large islands suggest that regional distributions should have the predictable shape that was first characterized by Brown and Nicoletto (1991). They should be somewhat right skewed and flatter than continental distributions, and they should become increasingly less skewed and flatter as the spatial scale decreases. This similarity in distribution shape implies there are underlying mechanisms that shape these distributions over ecological and evolutionary time. Moreover, it suggests that humans have manipulated the environment by default or design since they became a dominant species (e.g., Lyons et al. 2004; Donlan et al. 2005; Surovell et al. 2005; Donlan et al. 2006). Therefore,

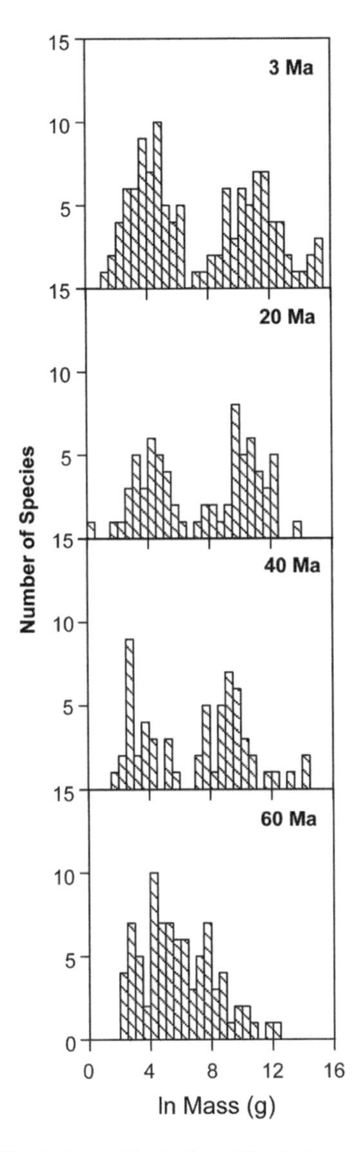


FIGURE 5.3. Frequency distributions of body size of North American mammals at four different points in time across the Cenozoic: 60, 40, 20, and 3 million years ago. Note that continental body size distributions are bimodal by 40 million years ago and remain that way until the end-Pleistocene extinctions. (Data courtesy of J. Alroy [Alroy 1998, 2000].)

present-day macroecological patterns must be interpreted with the realization that anthropogenic effects may have altered the patterns from their "natural" state. If we truly think that the repeatable macroecological patterns we are seeing have underlying ecological processes, then we must be sure that the patterns we are trying to explain are not unduly artificially altered.

### *Continental Patterns of Body Size and Range Size*

The relationship between body size and range size, for most groups examined, is a triangular relationship in which large-bodied species tend to have large ranges and small-bodied species have a large amount of variation in the shape of their geographic range (Brown 1995; Blackburn and Gaston 1996; Gaston and Blackburn 1996; Gaston 2003). Mammals in the present-day New World show this pattern (fig. 5.4, *left-hand panel*) both across all mammals and within the different orders (J. Madin, pers. comm.). Moreover, this pattern is consistent across the last 40,000 years in North America (fig. 5.4, *right-hand panels*). In fact, we see this messy, triangular relationship before and after the extinction of large mammals. Nonetheless, there are some differences. The body size/range size plots for the Holocene, Glacial, and Pre-Glacial time periods (fig. 5.4, *right-hand panels*) have some large-bodied mammals that had small ranges. However, examination of the identity of these points indicates that in each case, they are species that subsequently went extinct or are species whose distributions are truncated, as the majority of the range falls outside the United States (i.e., the range of the FAUNMAP data). For example, in the Pre-Glacial plots, the three largest mammals with small ranges are *Homotherium serum*, *Hydrochoerus holmesi*, and *Equus scotti*. Each of these species is now extinct. In the Glacial time periods, the large mammals with the smallest ranges are *H. holmesi* and *E. niobrarensis*. The other large-bodied species with small ranges are species like *Alces alces* that have a large portion of their current range outside of the range of the FAUNMAP data. Finally, in the Holocene the large-bodied species with small ranges tend to be species like the peccary and the mountain goat, whose ranges extend outside the United States.

The consistency over time of the relationship between body size and range size, the pattern that large-bodied mammals have large ranges and small-bodied mammals have great variation in range size, suggests that small mammals can be either coarse or fine grained in how they perceive

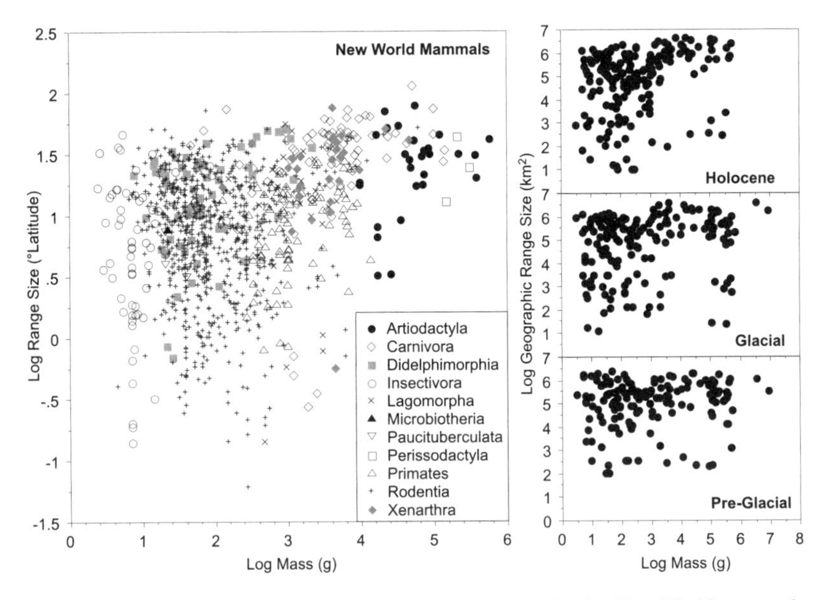


FIGURE 5.4. The relationship between body size and range size for New World mammals (*left-hand panel*) and for North American mammals at three time periods during the Quaternary: Holocene (last 10,000 years), Glacial (20,000–10,000 years ago), and Pre-Glacial (40,000–20,000 years ago). Data for New World mammals are coded according to ordinal affliation.

the environment, but that large mammals can only be coarse grained. This difference in how large- and small-bodied mammals perceive and use the environment could help explain the triangular shape of the relationship, because it explains why the smallest range size within a body size bin increases with body size. However, it does not explain the top constraint line, or lack thereof. Maximum range size in all body size bins is similar and seems to be related to available space. That is, at all body sizes, some species will have the maximum range size allowable given the available land area. Why are there small-bodied species that use a lot of space? One possibility is that there are constraints imposed by life history traits or phylogenetic history. However, this has yet to be explored across all mammals.

Interestingly, the overall triangular shape cannot be attributed to turnover in the ordinal identity across the *x* axis. The triangular-shaped relationship holds within orders as well as across all mammals (fig. 5.4, *left-hand panel*; J. Madin, pers. comm.), suggesting that mode of life influences the steepness of the relationship, but not the overall triangular shape. If the triangular shape were attributable simply to differences in how body size relates to range size in different orders of mammals, we would not expect to see the same pattern recapitulated within the different orders. Indeed, the fact that larger insectivores (fig. 5.4, *left-hand panel, open circles*) or larger artiodactyls (fig. 5.4, *left-hand panel, solid circles*) tend to have larger ranges and less variation in range size than smaller-bodied members of their order suggests that there is something about body size per se that influences the overall shape.

### *Community-Level Patterns of Body Size*

There are two main findings concerning body size distributions at the community level. The first is that they tend to be flatter than distributions at regional or continental scales (Brown and Nicolleto 1991). In general, this means that as the spatial scale of the analysis decreases, the number of species in the modal size classes is reduced. Brown and Nicolleto (1991) first showed this pattern for North American mammals and pointed out that this meant that beta diversity among communities was due to turnover in species from the modal size class (i.e., 100 g) and that smaller- and larger-bodied species were occurring in multiple communities. Studies that have extended these analyses to South America (Marquet and Cofre 1999; Bakker and Kelt 2000) and Africa (Kelt and Meyer 2009) have also found a flattening of the body size distribution with decreasing spatial scale.

The second main finding first recognized by Brown and Nicolleto (1991) is that the shapes of body size distributions change with spatial scale. They found that although North American continental body size distribution was unimodal and right skewed, local body size distributions were essentially uniform. The uniform shape of the distribution was consistent for all of the different habitats they examined. However, subsequent studies for South America (Marquet and Cofre 1999; Bakker and Kelt 2000) and Africa (Kelt and Meyer 2009) have not replicated these findings. Non–rain forest communities in South America showed the uniform distribution common to North America; however, rain forest communities had more unimodal and peaked distributions (Marquet and Cofre 1999; Bakker and Kelt 2000). The difference was attributed to the arboreal habitat available in rain forests and the additional medium-sized mammals that it could support. Interestingly, Africa shows a third pattern. In Africa, local body size distributions are flatter than continental body size distributions but never completely lose the bimodality found at larger spatial scales (Kelt and Meyer 2009). Kelt and Meyer (2009) attribute this difference to the second mode of large-bodied mammals still present in Africa and argue that any coevolutionary accommodations between early human hunters and large-bodied mammals that allowed the African fauna to escape anthropogenic extinction may also be responsible for the bimodality of local body size distributions. Interestingly, this predicts that other continents such as Eurasia that also retained their second mode of large-bodied mammals and avoided the end-Pleistocene extinction should show similar patterns to that of Africa.

ARE THE SHAPES OF BODY SIZE DISTRIBUTIONS AT THE COMMUNITY LEVEL CONSISTENT ACROSS SPACE? Our analyses of the moments of community body size distributions across space and time found significant differences in the shapes of community body size distributions among continents (fig. 5.5, tables 5.5, 5.6, and 5.7) and across time (fig. 5.6). However, there were notable exceptions. First, pairwise analyses of the median ln body mass of communities on different continents found significant differences among only three pairs of continents: Africa and North America, Eurasia and Australia, and Australia and South America (fig. 5.5, table 5.5). Africa had the highest median body size of its communities, followed by South America. As might be expected from the results of Kelt and Meyer (2009), the majority of the differences in median ln body size are driven by differences with Africa. Because Africa still retains its large-bodied fauna and the second mode in the continental body size distribution (fig. 5.1), African communities could have a higher median body size, because they contain more individuals of larger size. However, we did not find the expected similarity between Eurasian and African communities, despite the fact that the continental body size distributions of Africa and Eurasia are remarkably similar and both continents retain a significant portion of their large-bodied fauna (fig. 5.1). This may reflect the wider range of temperate latitudes in Eurasia than in Africa.

We found no difference in the skewness of the communities on different continents (fig. 5.5, table 5.6). The average skewness on each continent was close to zero, but the range of values was much greater on Australia, Eurasia, and South America (fig. 5.5). Clearly African communities did not attain higher median ln body sizes by having communities with radically different skewness. Indeed, the average skewness

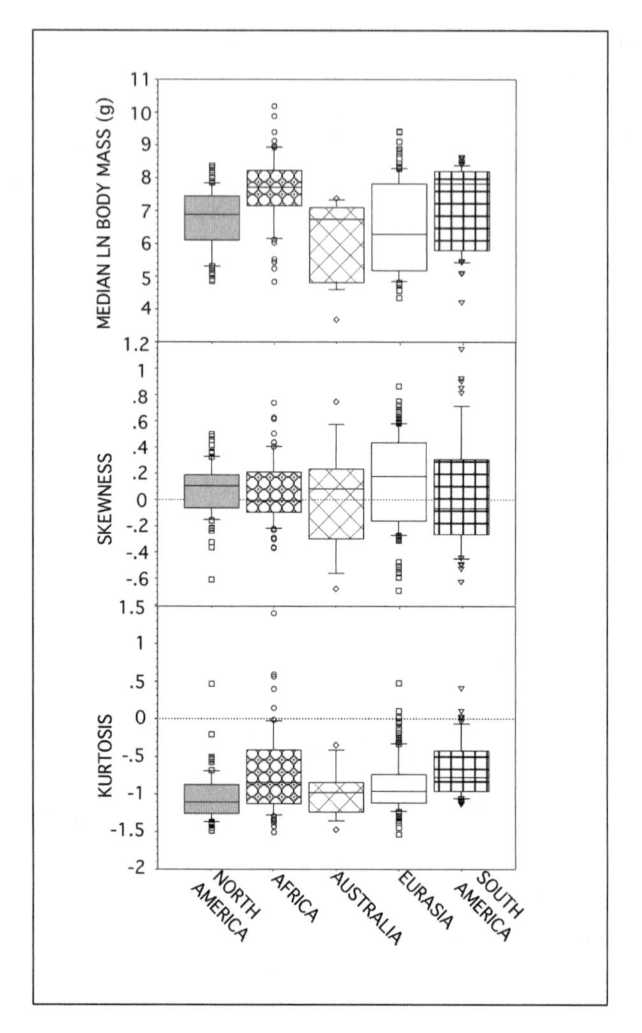


FIGURE 5.5. Box plots comparing the moments of body size distributions of modern mammal communities for five different continents, North America (*gray bar*), Africa (*circle hatching*), Australia (*x hatching*), Eurasia (*white bar*), and South America (*square hatching*). Each panel represents a different moment: median ln body mass, skewness, and kurtosis.

for African communities is approximately zero (fig. 5.5), consistent with the findings of bimodality reported by Kelt and Meyer (2009). Although it is not a significant difference, it is worth noting that Eurasia has the highest average skewness and that the values tend to be positive, indicating a right skew. Despite having a second mode of large-bodied mam-

TABLE 5.5 **Comparisons of ln Median Body Size for Body Size Distributions of Local Communities on Different Continents**

| | Africa | Eurasia | North America | South America | Australia |
|---|---|---|---|---|---|
| Africa | — | **<0.001** | **<0.001** | 0.05 | **<0.001** |
| Eurasia | 5.19 | | 0.38 | 0.007 | 0.23 |
| North America | −5.39 | 0.875 | | 0.01 | 0.03 |
| South America | 2.03 | −2.75 | −2.48 | | **0.005** |
| Australia | 4.71 | −1.20 | 2.20 | −2.95 | |

*Note*: Analyses were performed using *t*-tests. Values in the bottom triangle are test statistics, and values in the upper triangle are *p* values. Continent-specific generic averages were substituted for missing body sizes. A Bonferroni correction puts the alpha needed for significance at 0.005. Significant *p* values are highlighted in bold.

TABLE 5.6 **Comparisons of Skewness Values for Body Size Distributions of Local Communities on Different Continents**

| | Africa | Eurasia | North America | South America | Australia |
|---|---|---|---|---|---|
| Africa | — | 0.09 | 0.64 | 0.67 | 0.58 |
| Eurasia | −1.70 | | 0.11 | 0.07 | 0.17 |
| North America | 0.47 | −1.61 | | 0.40 | 0.34 |
| South America | 0.42 | 1.85 | 0.84 | | 0.88 |
| Australia | 0.56 | −1.40 | 0.96 | −0.15 | |

*Note*: Analyses were performed using *t*-tests. Values in the bottom triangle are test statistics, and values in the upper triangle are *p* values. Continent-specific generic averages were substituted for missing body sizes. A Bonferroni correction puts the alpha needed for significance at 0.005. There are no significant *p* values.

mals still extant (fig. 5.1), Eurasian communities are more likely to have a peak in smaller-bodied species and a long tail of large-bodied species rather than a second mode. Again, this is unlike Africa and suggests that simply having a continental biota that was less affected by the megafaunal extinction is not sufficient to explain African community body size distributions.

Analyses of the kurtosis values of the communities on different continents found significant differences largely between North and South America and the other continents (fig. 5.5, table 5.7). Specifically, the distribution of kurtosis values describing North American communities was significantly different from that of African, Eurasian, and South American communities. In addition to being significantly different from North America, South America was also significantly different from Eurasia. On all continents, the majority of communities had kurtosis values less than zero, indicating a relatively flat distribution. However, the variation in community kurtosis values on the different continents is infor-

TABLE 5.7 **Comparisons of Kurtosis Values for Body Size Distributions of Local Communities on Different Continents**

| | Africa | Eurasia | North America | South America | Australia |
|---|---|---|---|---|---|
| Africa | — | 0.02 | **<0.001** | 0.56 | 0.08 |
| Eurasia | 2.33 | | **<0.001** | **<0.001** | 0.23 |
| North America | −4.27 | −3.37 | | **<0.001** | 0.59 |
| South America | −0.59 | −3.56 | −6.20 | | **0.003** |
| Australia | 1.80 | −1.20 | −0.54 | −3.09 | |

*Note*: Analyses were performed using *t*-tests. Values in the bottom triangle are test statistics, and values in the upper triangle are *p* values. Continent-specific generic averages were substituted for missing body sizes. A Bonferroni correction puts the alpha needed for significance at 0.005. Significant *p* values are highlighted in bold.

mative (fig. 5.5). North America and Australia had kurtosis values that fell within a narrow range, with the average being close to −1, indicative of the flat distributions documented in Brown and Nicoletto (1991). Eurasia and South America also had kurtosis values that fell within a narrow range. However, the average for Eurasia was close to −1, whereas the average for South America was closer to −0.75. For both continents, the average values indicate a majority of communities with relatively flat distributions; however, the body size distributions of South American communities were slightly more peaked than those of Eurasian communities. This is consistent with the findings of more peaked distributions in South American rain forests (Marquet and Cofre 1999; Bakker and Kelt 2000). Interestingly, the range of values for Eurasian communities is much greater than for any other continent save Africa. This suggests that Eurasian community body size distributions run the gamut from highly bimodal to uniform to highly peaked.

African community body size distributions show the greatest range of kurtosis values and the highest average (fig. 5.5). Moreover, the communities with the absolute highest kurtosis values occur in Africa. This is in sharp contrast to the bimodality at all spatial scales reported by Kelt and Meyer (2009). The difference may be due to the area of the local communities used in the different studies and the greater number of local communities analyzed here. Our communities were constrained to be <10,000 km$^2$, whereas Kelt and Meyer's (2000) were not. Moreover, we analyzed a much larger number of local communities covering a wider range of habitats (57 vs. 14). Although some communities are peaked, it is not a consistent feature of African communities. Our analyses indicate that the majority of African communities have peaked distributions, but that they run the gamut from bimodal to uniform to highly peaked. Interestingly, in this respect African and Eurasian communities are similar.

The analyses presented here confirm and extend the findings previously reported concerning the shapes of community body size distributions. First, community body size distributions across the globe are flatter than either regional- or continental-level distributions. The flattening with spatial scale first noted by Brown and Nicoletto (1991) is a consistent characteristic of community body size distributions regardless of continent, clade composition or habitat and biome membership. Second, the shapes of the body size distributions are not consistently uniform as they are for North America. There are significant differences among and within all the continents. These differences are probably associated with differences in habitat type and clade composition. However, more research comparing body size distributions as a function of habitat type irrespective of continental association is necessary to evaluate that claim.

ARE THE SHAPES OF BODY SIZE DISTRIBUTIONS AT THE COMMUNITY LEVEL CONSISTENT ACROSS TIME? For each of the moments analyzed (median, skewness, and kurtosis), >60% of the values for the fossil communities fell within the same range as the values for modern North American communities. However, we did find significant differences in some of the moments in the different time periods. There were no significant differences in the median ln body size of late Pleistocene, Holocene, and modern North American communities (fig. 5.6, late Pleistocene vs. Holocene: $t = 1.91, p = 0.057$; late Pleistocene vs. modern: $t = -1.67, p = 0.10$; Holocene vs. modern: $t = -0.28, p = 0.78$). This is despite the fact that late Pleistocene communities still contained the megafauna that went extinct after the arrival of humans (Martin 1966, 1967, 1984; Martin and Klein 1984). The late Pleistocene communities have a greater variation in median ln body size than the more recent time periods, but this difference is not significant. It is possible that this is due to poorer sampling of the large-bodied species in the community. However, that is unlikely. Large-bodied species are more likely to be recorded than small-bodied species (Lyons and Smith 2006, 2010). In North America, approximately 80 species of large-bodied mammals went extinct at the end of the Pleistocene (Lyons et al. 2004). The greater range values of median ln body size may reflect greater competition for resources and community membership among large-bodied species when so many more of them are extant.

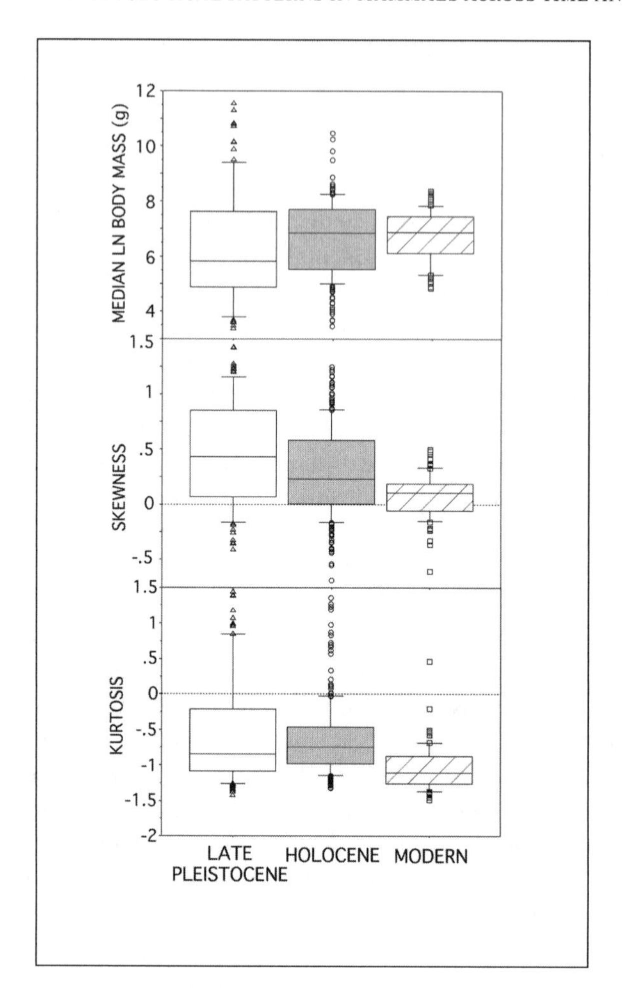


FIGURE 5.6. Box plots comparing the moments of body size distributions of late Pleistocene (*white bar*), Holocene (*gray bar*), and modern (*diagonal hatching*) mammal communities. Each panel represents a different moment: median ln body mass, skewness, and kurtosis.

The value of the median ln body size will be dependent on which and how many of the large-bodied species are present. If it is highly variable, median ln body size of communities should be as well.

Our analysis found significant differences in the skewness of communities for all pairwise combinations of time periods (fig. 5.6, late Pleistocene vs. Holocene: $t = -3.56$, $p < 0.001$; late Pleistocene: $t = 6.99$,

$p < 0.001$; Holocene vs. modern: $t = 4.68$, $p < 0.001$). Indeed, skewness decreased toward the present, with late Pleistocene communities significantly more right skewed than Holocene communities, which were in turn significantly more skewed than modern communities. However, what seems to really be driving the pattern is the much greater range of skewness values describing the fossil communities (fig. 5.6). There are two possible explanations for this. The first is that it is a sampling problem and the use of fossil data is introducing noise into the data that is increasing variation. To some degree, this is probably the case. The second possibility is that the climate change and its affect on communities is affecting community body size distributions. The climate change associated with glaciation had significant effects on the distributions of mammalian species (Graham 1984, 1986; Graham and Mead 1987; Graham et al. 1996; Lyons 2003) and there was considerable turnover in community composition (Lyons 2005). Moreover, late Pleistocene communities were more diverse than Holocene and modern communities (Graham et al. 1996). Although it has been less well studied, it is possible that the rapid climate flickers during the Holocene also affected community composition and diversity. These differences in other aspects of community structure may have translated into differences in community body size distributions as well.

With respect to the kurtosis, we found that the fossil communities in the late Pleistocene and Holocene were significantly different from the modern communities, but not significantly different from each other (fig. 5.6, late Pleistocene vs. Holocene: $t = -1.67$, $p = 0.10$; late Pleistocene vs. modern: $t = 5.56$, $p < 0.001$; Holocene vs. modern: $t = 6.59$, $p < 0.001$). Moreover, the average kurtosis for both Holocene and late Pleistocene communities was higher than for modern communities, suggesting at least some communities with more peaked and less uniform distributions. Indeed, the smallest kurtosis values in each time period were similar, but the late Pleistocene and Holocene contained communities with much higher kurtosis values than found in modern communities (fig. 5.6). The differences in kurtosis probably have the same explanation as differences in skewness and are a result of noise introduced by the fossil data and real differences resulting from differences in community structure of North American communities over the last 40,000 years.

Although the analyses presented here document significant differences in some of the moments of the body size distributions of mammalian communities in North America over the last 40,000 years, it is worth emphasizing that the majority of communities (>60%) had values for each of the moments that fell within the range of modern North America. This suggests that the majority of communities had the flat distributions documented by Brown and Nicoletto (1991) and that relatively flat, uniform distributions are common to temperate North America. However, the greater range of values suggests that the changes in climate and differences in community composition and diversity did play a role in the shapes of body size distributions. Moreover, it implies that, as with differences on the different continents, habitat type is likely to be an important predictor in the shapes of these distributions across both time and space.

## Summary and Conclusions

Our comprehensive analysis of the macroecology of body size in mammals across space and time shows that there are gross similarities in these macroecological patterns across land masses and throughout evolutionary time. Although there are unique aspects to the patterns on individual continents, the shapes of body size distributions and the moments of the distributions are remarkably similar despite their different evolutionary and geological histories (figs. 5.1 and 5.2). Moreover, the overall shape of the continental body size distribution seems to have developed by 40 million years ago and remained consistent until the extinction of the megafauna at the end of the Pleistocene (fig. 5.3). At a local scale, studies have shown that the shape of the body size distribution differs depending on your continent and biome (Brown and Nicoletto 1991; Marquet and Cofre 1999; Bakker and Kelt 2000; Kelt and Meyer 2009). Our analysis confirms that there are differences among the three continents studied and extends the finding to Australia and Eurasia (fig. 5.5). However, we also find a similarity to the patterns across the late Pleistocene, suggesting that the habitat- and biome-level factors that shape community-level body size distributions have done so consistently across time (fig 5.6). Finally, we find that other macroecological patterns, such as the relationship between body size and range size, are recoverable in the fossil record and that any discrepancies between modern patterns and fossil patterns can be attributed to the anthropogenic extinction of the late Pleistocene megafauna.

## References

Alroy, J. 1998. "Cope's rule and the dynamics of body mass evolution in North American fossil mammals." *Science* 280 (5364): 731–734.

———. 2000. "New methods for quantifying macroevolutionary patterns and processes." *Paleobiology* 26 (4): 707–733.

Bakker, V. J., and D. A. Kelt. 2000. "Scale-dependent patterns in body size distributions of Neotropical mammals." *Ecology* 81 (12): 3530–3547.

Blackburn, T. M., and K. J. Gaston. 1996. "Spatial patterns in the geographic range sizes of bird species in the New World." *Philosophical Transactions of the Royal Society B: Biological Sciences* 351 (1342): 897–912.

Bonaccorso, F. J. 1998 *Bats of Papua New Guinea*. Conservation International Tropical Field Guide Series 2. Washington DC: Conservation International.

Brown, J. H. 1995. *Macroecology*. Chicago: University of Chicago Press.

Brown, J. H., P. A. Marquet, and M. L. Taper. 1993. "Evolution of body size: Consequences of an energetic definition of fitness." *American Naturalist* 142 (4): 573–584.

Brown, J. H., and P. F. Nicoletto. 1991. "Spatial scaling of species composition: Body masses of North American land mammals." *American Naturalist* 138 (6): 1478–1512.

Calder, W. A., III. 1984. *Size, function, and life history*. Cambridge, MA: Harvard University Press.

Donlan, C. J., J. Berger, C. E. Bock, D. A. Burney, J. A. Estes, D. Foreman, P. S. Martin, G. W. Roemer, F. A. Smith, M. E. Soulé, and H. W. Greene. 2005. "Rewilding North America." *Nature* 436:913–914.

———. 2006. "Pleistocene rewilding: An optimistic agenda for twenty-first century conservation." *American Naturalist* 168:660–681.

FAUNMAP Working Group. 1994. *A database documenting late Quaternary distributions of mammal species in the United States*. Vol. 1. Illinois State Museum Scientific Papers, vol. 25. Springfield: Illinois State Museum.

Flannery, T. 1995. *Mammals of New Guinea*. Rev. and updated ed. Ithaca: Cornell Unversity Press.

Garbutt, N. 1999. *Mammals of Madagascar*. New Haven: Yale University Press.

Gaston, K. J. 2003. *The structure and dynamics of geographic ranges*. Oxford: Oxford University Press.

Gaston, K. J., and T. M. Blackburn. 1996. "Range size body size relationships: Evidence of scale dependence." *Oikos* 75 (3): 479–485.

Graham, R. W. 1984. "Paleoenvironmental implications of the Quaternary distribution of the eastern chipmunk (Tamias-Striatus) in central Texas." *Quaternary Research* 21 (1): 111–114.

———. 1986. "Response of mammalian communities to environmental changes during the late Quaternary." In *Community Ecology*, edited by J. Diamond and T. J. Case, 300–313. New York: Harper and Row.

Graham, R. W., E. L. Lundelius, M. A. Graham, E. K. Schroeder, R. S. Toomey, E. Anderson, A. D. Barnosky, J. A. Burns, C. S. Churcher, D. K. Grayson, R. D. Guthrie, C. R. Harington, G. T. Jefferson, L. D. Martin, H. G. McDonald, R. E. Morlan, H. A. Semken, S. D. Webb, L. Werdelin, and M. C. Wilson. 1996. "Spatial response of mammals to late Quaternary environmental fluctuations." *Science* 272 (5268): 1601–1606.

Graham, R. W., and J. I. Mead. 1987. "Environmental fluctuations and evolution of mammalian faunas during the last deglaciation in North America." In *North American and adjacent oceans during the last deglaciation*, edited by W. F. Ruddiman and H. E. Wright, Jr., 372–402. Boulder, CO: Geological Society of America.

Grainger, M., E. Gunn, and D. Watts. 1987. *Tasmanian mammals: A field guide.* Hobart: Tasmanian Conservation Trust.

Kelt, D. A., and M. D. Meyer. 2009. "Body size frequency distributions in African mammals are bimodal at all spatial scales." *Global Ecology and Biogeography* 18 (1): 19–29.

Lyons, S. K. 2003. "A quantitative assessment of the range shifts of Pleistocene mammals." *Journal of Mammalogy* 84 (2): 385–402.

———. 2005. "A quantitative model for assessing community dynamics of Pleistocene mammals." *American Naturalist* 165 (6): E168–E185.

———. 2007. "The relationship between environmental variables and mammalian body size distributions over the space and time." Paper read at the annual meetings of the Ecological Society of America, San Jose, CA.

Lyons, S. K., and F. A. Smith. 2006. "Assessing biases in the mammalian fossil record using late Pleistocene mammals from North America." *Geological Society of America Abstracts with Programs* 38: 307.

———. 2010. "Using a macroecological approach to study geographic range, abundance, and body size in the fossil record." In *Quantitative methods in paleobiology*, edited by J. Alroy and G. Hunt, Paleontological Society Papers 16, 117–141.

Lyons, S. K., F. A. Smith, and J. H. Brown. 2004. "Of mice, mastodons and men: Human-mediated extinctions on four continents." *Evolutionary Ecology Research* 6 (3): 339–358.

MacPhee, R. D. E., and C Flemming. 1999. "Requiem aeternam: The last five hundred years of mammalian species extinctions." In *Extinctions in near time: Causes, contexts, and consequences*, edited by R. D. E. MacPhee, 333–372. New York: Kluwer Academic/Plenum.

Madin, J. S., and S. K. Lyons. 2005. "Incomplete sampling of geographic ranges weakens or reverses the positive relationship between an animal species' geographic range size and its body size " *Evolutionary Ecology Research* 7:607–617.

Marquet, P. A., and H. Cofre. 1999. "Large temporal and spatial scales in the structure of mammalian assemblages in South America: A macroecological approach." *Oikos* 85 (2): 299–309.

Martin, P. S. 1966. "Africa and Pleistocene overkill." *Nature* 212 (5060): 339–342.

——. 1967. "Prehistoric overkill." In *Pleistocene extinctions: The search for a cause*, edited by P. S. Martin and H. E. Wright Jr., 75–120. New Haven: Yale University Press.

——. 1984. "Prehistoric overkill: The global model." In *Quaternary extinctions: A prehistoric revolution*, edited by P. S. Martin and R. G. Klein, 354–403. Tucson: University of Arizona Press.

Martin, P. S., and R. G. Klein. 1984. *Quaternary extinctions: A prehistoric revolution*. Tucson: University of Arizona Press.

Martin, P. S., and D. W. Steadman. 1999. "Prehistoric extinctions on islands and continents." In *Extinctions in near time: Causes, contexts, and consequences*, edited by R. D. E. MacPhee, 17–55. New York: Kluwer Academic/Plenum.

Niklas, K. J. 1994. *Plant allometry: The scaling of form and process*. Chicago: University of Chicago Press.

Patterson, B. D., G. Geballos, W. Sechrest, M. Toghelli, G. T. Brooks, L. Luna, P. Ortega, I. Salazar, and B. E. Young. 2004. *Digital distribution maps of the mammals of the Western Hemisphere*. Version 1.0. Nature Serve 2003. Cited February 2004. Available from http://www.natureserve.org/getData/mammalmaps.jsp.

Peters, R. H. 1983. *The ecological implications of body size*. Cambridge: Cambridge University Press.

Smith, F. A., J. H. Brown, J. P. Haskell, S. K. Lyons, J. Alroy, E. L. Charnov, T. Dayan, B. J. Enquist, S. K. M. Ernest, E. A. Hadly, K. E. Jones, D. M. Kaufman, P. A. Marquet, B. A. Maurer, K. J. Niklas, W. P. Porter, B. Tiffney, and M. R. Willig. 2004. "Similarity of mammalian body size across the taxonomic hierarchy and across space and time." *American Naturalist* 163 (5): 672–691.

Smith, F. A., S. K. Lyons, S. K. M. Ernest, and J. H. Brown. 2008. "Macroecology: More than the division of food and space among species on continents." *Progress in Physical Geography* 32 (2): 115–138.

Smith, F. A., S. K. Lyons, S. K. M. Ernest, K. E. Jones, D. M. Kaufman, T. Dayan, P. A. Marquet, J. H. Brown, and J. P. Haskell. 2003. "Body mass of late Quaternary mammals." *Ecology* 84 (12): 3403–3403.

Strahan, R. 1995. *Mammals of Australia.* . Washington, DC: Smithsonian Institution Press.

Stuart, A. J. 1999. "Late Pleistocene megafaunal extinctions: A European perspective." In *Extinctions in near time: Causes, contexts, and consequences*, edited by R. D. E. MacPhee, 257–270. New York: Kluwer Academic/Plenum.

Surovell, T., N. Waguespack, and P. J. Brantingham. 2005. "Global archaeological evidence for proboscidean overkill." *Proceedings of the National Academy of Sciences of the United States of America* 102 (17): 6231–6236.

Wilson, D. E., and D. M. Reeder. 1993. *Mammal species of the world: A taxonomic and geographic reference.* 2nd ed. Washington, D.C: Smithsonian Institution.

PART II

# Mechanisms and Consequences Underlying Body Size Distributional Patterns

CHAPTER SIX

# Using Size Distributions to Understand the Role of Body Size in Mammalian Community Assembly

S. K. Morgan Ernest

Of the broad array of organismal traits that can be quantified, size is one of the few that has been widely studied across a range of taxonomic groups. Considered independently of other species traits, it is not immediately clear why size should be of such general interest. Why is it important whether one mammal is three times as large as another? And why would size be equally relevant in plants, fish, birds, plankton, and almost every other taxonomic group on the planet? It is precisely because size is not independent of other species' characteristics that it is such an important ecological trait to study. A quick examination of the classic books on allometry (Peters 1983; Calder 1984; Schmidt-Nielsen 1984; Niklas 1994) demonstrates just how many traits are correlated with organism size and for just how many taxa this is true. Everything from physiological processes (e.g., metabolic rate, gut fermentation times, water flux, and nutritional requirements) to individual requirements (e.g., home range size) to population variables (e.g., population growth, population production, and population size) is correlated with organism size. The fact that size is highly integrated with so many important traits and that this is true for almost all lineages of life is one of the great wonders of biology.

While the importance of size has long been appreciated in the study of morphology and physiology (e.g., Thompson 1917; Kleiber 1932; Brody 1945), its relevance to community ecology has only more recently been recognized. There are a number of reasons why the size of organisms should influence how they interact with each other in a community context. Many of the traits referenced above are related to the resource requirements of an organism, how it processes those resources, and how those resources are allocated to growth and reproduction. Some traits, such as nutrient uptake and excretion rates, also have important ecosystem-level impacts (e.g., Allen et al. 2005; Torres and Vanni 2007) which can affect nutrient availability to a wide range of species. These traits therefore constitute important axes of a species' niche and are thus important for niche-based coexistence among species within a community. In addition to the mediation of species interactions through niche traits, size is often important in determining how individuals of different species interact when they come into contact with one another. Many studies have shown that body size plays an important role in predator-prey relationships by constraining the range of prey a predator can effectively use and by determining optimal prey size (Rosenzweig 1966; Schoener 1968; Peters 1983; Gittleman 1985; Phillips and Shine 2004; e.g., Barnes et al. 2010). In competition, larger species are often behaviorally dominant over smaller species, allowing larger species to win direct contests over resources (e.g., Blaustein and Risser 1976; Hersteinsson and MacDonald 1992; Rychlik and Zwolak 2006). This inextricable relationship among size, niche axes, and species interactions clearly makes size an important aspect of community ecology.

In the mid-twentieth century three seminal publications (Hutchinson and MacArthur 1959; Yoda et al. 1963; Sheldon and Parsons 1967) recognized that body size is important for understanding assemblages of species and that examining distributions of body size provides a useful approach for studying these processes. Sheldon and Parsons (1967) and Yoda et al. (1963), studying aquatic and forest ecosystems, respectively, both focused on the distribution of the sizes of individuals (i.e., the size spectra and self-thinning relationships; see fig. 6.1). In contrast, Hutchinson and MacArthur (1959) focused on patterns related to the distribution of the average sizes of the component species in an assemblage (i.e., the species size distribution; sensu White et al. 2007; see fig. 6.1). Since the publication of these seminal works, the study of community-level body size patterns has expanded rapidly, with each of these papers spawning productive, but largely distinct, bodies of literature. This resulted in separate research areas that focus on different levels of pattern (i.e., individual vs. species) and therefore on different aspects of the role that body size plays in structuring communities.

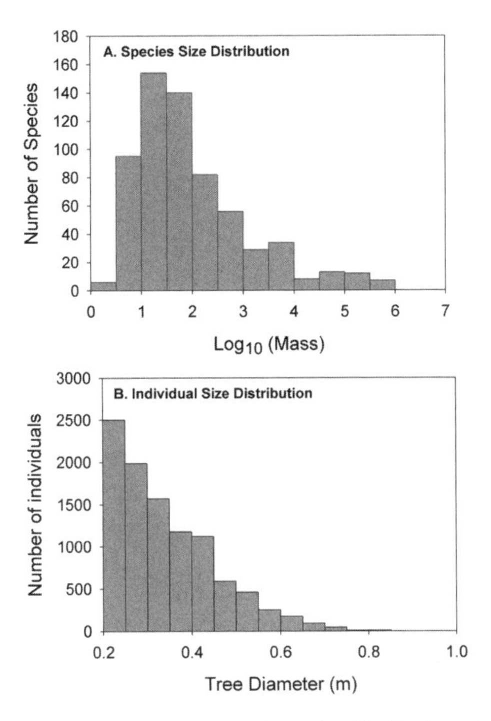


FIGURE 6.1. Examples of species size and individual size distributions. *A*, Species size distribution for extant mammals of North America. Data from Smith et al. (2003). *B*, Self-thinning relationship for trees at Corcieux, France. Data from de Liocourt (1898).

While the influence of body size on community ecology has been broadly studied across taxonomic groups, size appears to be particularly important in the assembly of mammal communities. However, most of the research in mammal communities has been developed in the tradition of Hutchinson and MacArthur (1959) and thus focused almost entirely on species-level patterns. In this chapter I will (1) examine the evidence demonstrating the role body size plays in mammal communities, (2) discuss the important insights that can be obtained by broadening the mammalian research agenda to include the use of individual-level body size distributions, and (3) examine the potential benefits of simultaneously exploring and comparing the relationships between individual and species-level body size distributions using data from nine small mammal communities.

## Species Body Size as a Structuring Factor in Mammal Communities

In 1959, Hutchinson and MacArthur observed that "the environment does not provide adequate room for a very large number of species of large animals while there is much more room for an abundance of smaller species." They based this statement on the idea that landscapes are made up of a mosaic of distinct "elements," where elements can consist of different resource or habitat types, and that each unique combination of elements constitutes a different niche. They showed that for a landscape containing five different elements distributed randomly, there are few niches available to species specializing on one element or to species requiring many elements, but there are a larger number of niches available to species of intermediate requirements (fig 6.2). To make predictions about the distribution of body sizes they combined this basic framework with three additional assumptions: (1) the number of elements a species uses is related to the size of the organism, with larger species requiring more elements, (2) only one species can be supported per niche, due to competitive exclusion, and (3) competition will be strongest between species of similar size and thus create spacing along the body size axis among species with similar niches. From this foundation, their model predicts a right-skewed lognormal distribution of species sizes. Hutchinson and MacArthur tested their model using data on five different taxa. Of these, only one group—mammals—matched their prediction. However, it has since been shown that many taxa exhibit the predicted right-skewed logarithmic species size distribution (plants: Van Valen 1973; beetles: Morse et al. 1988; e.g., fish: Brown et al. 1993; birds: Gaston and Blackburn 1995; bivalves: Jablonski and Raup 1995), and it has become increasingly clear that this pattern, because of its generality, provides important insights into the diversification and assembly of many different taxonomic groups.

Perhaps because the mammal data validated Hutchinson and MacArthur's (1959) predictions, mammals are the taxa where the model's core ideas have best been explored. For a group such as mammals, where species can differ in size by many orders of magnitude, the idea that animals of different size would perceive the same landscape very differently is clearly justified; a mouse will simply be incapable of using a landscape at the same scale as an elephant. Furthermore, physiological constraints—which

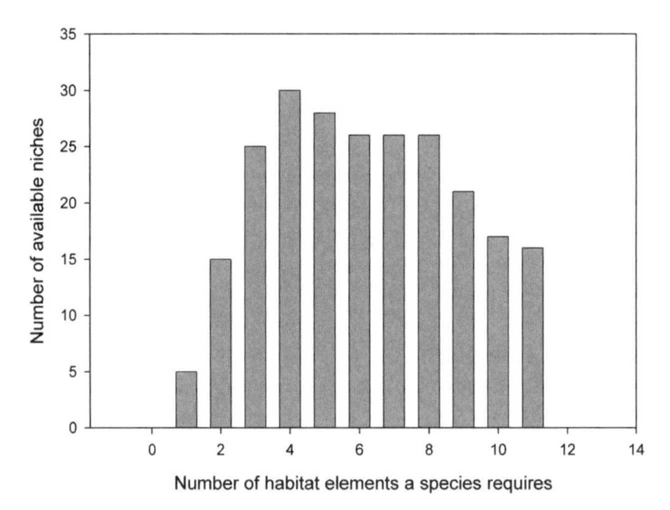


FIGURE 6.2. The distribution of niche availability with respect to the number of elements a species requires. Hutchinson and MacArthur (1959) equate the number of elements a species needs with its size (i.e., small species can specialize on just one element, while larger species require multiple elements). Assuming this correspondence between size and specialization, the number of niches should be higher for species of between the smallest and intermediate size classes than for the smallest and largest species. Data from Hutchinson and MacArthur (1959).

determine the amounts and quality of forage on which an individual can survive—differ dramatically between mammals of different size (Owen-Smith 1988; Belovsky 1997; McNab 2002). Even if a mouse could somehow operate at the same spatial scales as an elephant, it could not survive on the poorer quality of forage that an elephant consumes (Owen-Smith 1988).

There is also strong empirical evidence supporting the Hutchinsonian idea that species of similar size and resource requirement should exhibit strong competition and rarely coexist (Hutchinson 1959; Hutchinson and MacArthur 1959). At local scales (i.e., communities), mammalian species size distributions are generally uniform in structure—no body size class is more represented than any other (Brown and Nicoletto 1991). This pattern at local scales appears to emerge from strong competition between species of similar size, as evidenced by studies showing distinct spacing along the body size axis among coexisting species (Brown 1975; Bowers and Brown 1982). The uniform species size distribution of many communities probably results from limited niches and intense competition within a body size class. This idea is further supported by studies on species size distributions for tropical mammalian communities even though these distributions are often not uniform but exhibit an internal mode in the distribution that corresponds to primates (Marquet and Cofre 1999; Bakker and Kelt 2000). This suggests that the increased habitat complexity of tropical forests supports a unique life history and allows additional species of a certain size to coexist because of the addition of a new niche axis, vertical distribution within the canopy. As mammalian communities are aggregated by going up in scale (i.e., combining communities in the same biome or continent), the species size distribution shifts from uniform to the classic right-skewed lognormal (Brown and Nicoletto 1991). This shift in shape is typically interpreted as support for both the strong competition that prevents species of similar size from coexisting at local scales and the ability of smaller intermediate-sized species to specialize on a higher diversity of niches, thus allowing more of them to exist at broad spatial scales (Brown and Nicoletto 1991; Ritchie 1998). This strong role of body size, and in particular the strong competition between species of similar size, is in contrast to some other taxa, such as birds, where similar-sized species using similar resources can readily coexist in local communities (Thibault et al. 2011), sometimes even foraging within the same tree (MacArthur 1958). Whether or not many of Hutchinson and MacArthur's core concepts are relevant to other taxa, their ideas have found strong empirical support in mammals and have provided important insights into how mammalian assemblages are structured.

## Insights Gained from Individual Size Distributions

In contrast to the species size distribution, distributions of individual body size within communities, irrespective of species identity, have not been well studied in mammalian communities. However, these individual size distributions have been of great interest in other groups, the most well studied being the biomass spectrum in aquatic communities and the self-thinning rule in forests. Both patterns examine the distribution of size at the individual level, eschewing the traditional community ecology focus on species and focusing instead on the functional importance of size. This interest in size arises in part because of shared interests in aquatic and forest ecology in predicting sustained ecosystem productivity and a realization that biomass production is a function of size (de Liocourt 1898; Yoda et al. 1963; Sprules and Munawar 1986). These similarities have resulted in some similar insights regarding how size structures aquatic and forest ecosystems. First, body-size-mediated biotic interactions among individuals are important for understanding the shape of the individual size distribution. For example, self-thinning distributions are often modeled based on individual-level competition for limiting resources within and between size classes, instead of between populations of species (e.g., Muller-Landau et al. 2006; West et al. 2009). Second, individual-level processes such as resource acquisition and allocation are intimately tied to community and ecosystem patterns of productivity. For example, biomass spectrum models are often built up from an individual metabolic basis to explain both the biomass productivity of different size classes and the impacts of consumption by some size classes on other size classes (Kerr and Dickie 2001).

One of the novel insights that can be gained from the individual size distribution is the linkage between size and ecosystem processes. Because of the relationship between size and rates of resource use in many taxa (e.g., Kleiber 1932; Hemmingsen 1960; Savage et al. 2004), the individual size distribution reflects how resources are processed by an assemblage and which size classes are playing dominant roles in energy flow through ecosystems. Because of the relationship between size and biomass production (e.g., Ernest et al. 2003), the individual size distribution also can be used to understand how biomass is partitioned in ecosystems and understand how rates of biomass production shift with changes in the individual size distribution (Kerr and Dickie 2001). As a consequence, the individual size distribution yields a very different perspective on community and ecosystem structure and function from the species size distribution, which focuses solely on average species size and is not directly convertible to currencies relevant to ecosystem processes.

In contrast to the well-studied nature of individual size distributions in aquatic systems and forests, only a handful of studies have attempted to empirically examine these patterns in terrestrial animal communities (Schoener and Janzen 1968; Griffiths 1986; Morse et al. 1988; Siemann et al. 1996, 1999; Thibault et al. 2011). Despite the fact that mammalogists typically collect exactly the kind of data needed to construct individual size distributions—the weight of every individual they catch—only two studies of which I am aware have examined this relationship in mammals (Griffiths 1986; Ernest 2005). Our understanding of the influence of size on mammalian communities is hampered by this lack of investigation of the individual size distribution. This may be particularly important because at least one study has shown that shifts in the individual size distribution in mammal communities can have profound consequences for community properties such as the number of individuals and the total biomass an assemblage can support; the same amount of resource can support many small individuals or a few larger individuals (White et al. 2004; Ernest et al. 2008). Understanding why mammalian size distributions shift and the ecosystem consequences for these shifts will require an increased empirical emphasis on this central ecological pattern.

As one might expect given the limited empirical documentation on the individual size distribution in mammals, there is also little theoretical understanding of what processes might control its shape and even less ability to predict how it might shift in response to changes in the environment. One of the theoretical models available for understanding the individual size distribution in terrestrial consumer assemblages is the textural discontinuity hypothesis (Holling 1992). The textural discontinuity hypothesis proposes that only a few processes control how resources are available in any ecosystem. These processes operate at distinct spatial scales and temporal frequencies, causing resources to be available patchily across scales (Holling 1992). Because body size determines the spatial and temporal scales at which organisms operate (Peters 1983; Calder 1984; Schmidt-Nielsen 1984), resources will be available in a discontinuous manner with regard to body size. This is an idea similar to the fractal distribution of resources that is often posited for structuring the species size distribution (e.g., Ritchie and Olff 1999), but with the added complexity that there are actually sizes of organisms that cannot be supported in an ecosystem because resources are not sufficiently available to those size classes. Because of that, the textural discontinuity hypothesis predicts that the species size distribution should exhibit clumps of species along the body size axis, with gaps occurring between these clumps (Allen et al. 2006). While the predictions of this model are typically evaluated based on the species size distribution, one of the intriguing aspects of this hypothesis is that it implicitly integrates both the species size distribution and the individual size distribution. If resources are patchily available with respect to body size, then individuals, as well as species, should be patchily distributed along the size axis. Therefore, the textural discontinuity hypothesis also predicts multimodal individual size distributions. Two studies documenting individual size distributions in terrestrial animal communities suggest that many animal communities may indeed exhibit multimodality in this distribution (Griffiths 1986; Thibault et al. 2011).

## Comparing Individual and Species-Based Size Distributions in Small Mammal Communities

While the textural discontinuity hypothesis provides an avenue for linking individual and species-based size distributions, whether its predictions match empirical reality has generally been untested. In addition to the relative paucity of individual size distributions for terrestrial animal communities, determining whether gaps in size distributions exist has been controversial (e.g., Manly 1996; Siemann and Brown 1999). An alternative approach to identifying gaps is to identify clusters of species (or individuals) through the identification of modes in size distributions (Manly 1996; Siemann and Brown 1999; Ernest 2005). One advantage of focusing on modes instead of gaps is that it refocuses the question in a broader and more tractable manner. Focusing on gaps is in effect asking the question whether there are body sizes that cannot be supported in a given ecosystem (Thibault et al. 2011). The difficulty of proving a negative is what has generated much of the controversy on this topic (i.e., it is practically impossible to distinguish between a low probability and a zero probability). By focusing on the presence of internal modes in the distribution, the question shifts to the relationship between resource availability and body size aggregation: are some body sizes better supported than others in a given ecosystem?

One of the clear predictions that emerges from this restructuring of how we assess the textural discontinuity hypothesis is that if resources are disproportionately available to certain body sizes, then not only should species sizes be clustered along the body size axis, but so should individuals. If there are enough resources to support more populations of species of a specific mass range, then the aggregation of those species should also result in more individuals aggregated along that portion of the size axis. This is similar in concept to the more-individuals hypothesis of species energy theory (e.g., Wright 1983); more resources can support more individuals and thus more species can be supported, because more populations can be above a critical extinction threshold for population size (Siemann et al. 1996, 1999).

To date, only one paper has assessed the prediction from the textural discontinuity hypothesis that the individual size distribution and the species size distribution should exhibit modes at the same locations along the size axis. In Ernest (2005), data was obtained for nine small-mammal communities in the continental United States. These investigations consisted of intensive surveys of small mammals under 500 grams. Most of the sites were located in the southwestern United States, though sites from the mountains of Oregon and Colorado (one each) were also included. For all these communities, every individual was weighed, providing detailed individual-level size data that could be used to generate individual size distributions, and the individuals were averaged by species to create species size distributions. Species size distributions were smoothed using kernel density estimation to facilitate mode location in distributions with low sample sizes. To assess whether species were aggregated along the size axis where the most resources were being used by the community, the individual size distribution was converted to a body size energy distribution (i.e., the distribution of energy use with respect to size). This conversion was accomplished by using a standard allometric relationship between mass and metabolic rate to calculate energy requirements for each individual: Metabolic Rate = $b$*Mass$^{3/4}$, where $b$ is a scaling constant (e.g., Kleiber 1932; West et al. 1997). Individual energy requirements were summed within body size classes to yield the energy use profile of the community with respect to body size. For each community, the body size energy distribution and the species size distribution were compared to assess whether the distributions exhibited correspondence in their structures, as would be predicted from the textural discontinuity hypothesis.

While both the body size energy and species size distributions exhibited internal structures consistent with modes, the locations of these structures did not generally match (fig 6.3). Modes of high energy flux along the body size axis did not generally match up with locations where clumps of species were occurring. In fact, across all nine communities, most modes were dominated by a single species and clumps of species were located in areas of the body size energy distribution where very little energy was being fluxed. In contrast to the predictions of the textural discontinuity hypothesis, this seemed to indicate that species were not clustering where resources were abundant, but where competitively dominant species were absent. However, what remained unclear from that study was what process was causing the location of modes along the

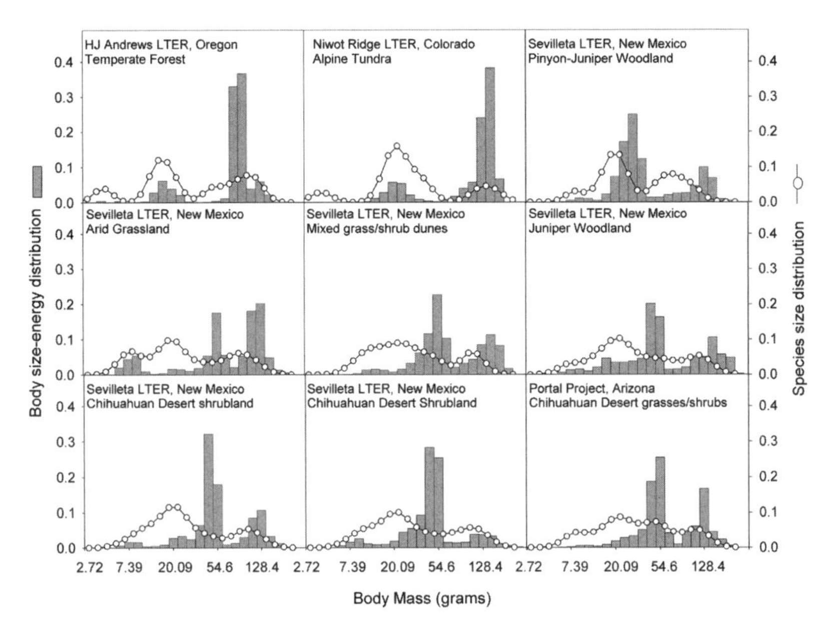


FIGURE 6.3. Comparison of individual size distributions (i.e., body size energy distribution) and species size distributions for nine small mammal communities. Adapted from Ernest (2005).

body size energy axis. Two possibilities were proposed: (1) locations of energy modes are idiosyncratic results of community assembly and determined solely by the species composition of the community, or (2) resource availability is determined by the ecosystem structure, and thus the location of modes along the body size axis should be relatively insensitive to the turnover of species as long as habitat structure has not changed. However, since Ernest (2005) there have been no follow-up papers on this question.

One approach to addressing this lingering question from Ernest (2005) is to examine how the structure of the species size distribution and the body size energy distribution change across space and through time. Sites with similar habitat structure should, in theory, be more similar in the shapes of their distributions than sites of dissimilar habitat if the ecosystem is determining which body sizes will be successfully supported. However, if the structures of these distributions are determined solely by which species are present in a community, then these distributions should change in response to differences in species composition.

Using the same data used in Ernest (2005), we can assess whether the shape of the distributions changes as sites are separated by greater distances and increase in their difference in species composition. We can also leverage the twenty years of data collection at one of the sites, the Portal Project, to assess how the species size and body size energy distributions respond to shifts in species composition through time. The relationships between species composition and the species size and body size energy distributions were determined by comparing all possible pairs of sites and calculating the differences in their species composition, species size, and body size energy distributions. Euclidean distance (ED) was used to determine the difference in species composition between two sites. ED measures the distance in multidimensional space between two samples, with large distances between samples corresponding to large differences in species composition (Collins 2000; Ernest and Brown 2001). The ED equation is

$$ED = \left( \sum_{i=1}^{S} (x_{ai} - x_{bi})^2 \right)^{1/2},$$

where $x$ is the relative abundance of species $i$ in study area $a$ or $b$. A simple metric for calculating the overlap between two distributions was used to compare the species size or body size energy distributions between two sites. This distribution overlap index (DOI) is

$$DOI = \sum_{k=1}^{p} \left| ( y_{ak} - y_{bk} ) \right|,$$

where $y$ is the value for bin $k$ for study site $a$ or $b$. Because all distributions were normalized to sum to 1, this index has a firm upper and lower limit, where 0 indicates complete overlap and 2 indicates complete non-overlap of the distributions.

While sites exhibited the same general multimodal shapes in their body size energy distributions, direct comparison of sites revealed that the distributions of all sites were significantly different from one another (all pairwise comparisons of DOI: all $p > 0.001$; Ernest 2005) and that sites that were located closer together had greater overlap in their energy distributions (fig 6.4). Some of the relationship between overlap in the energy distribution and spatial distance can potentially be explained by similarity in species composition (fig 6.4), but because of the topographic

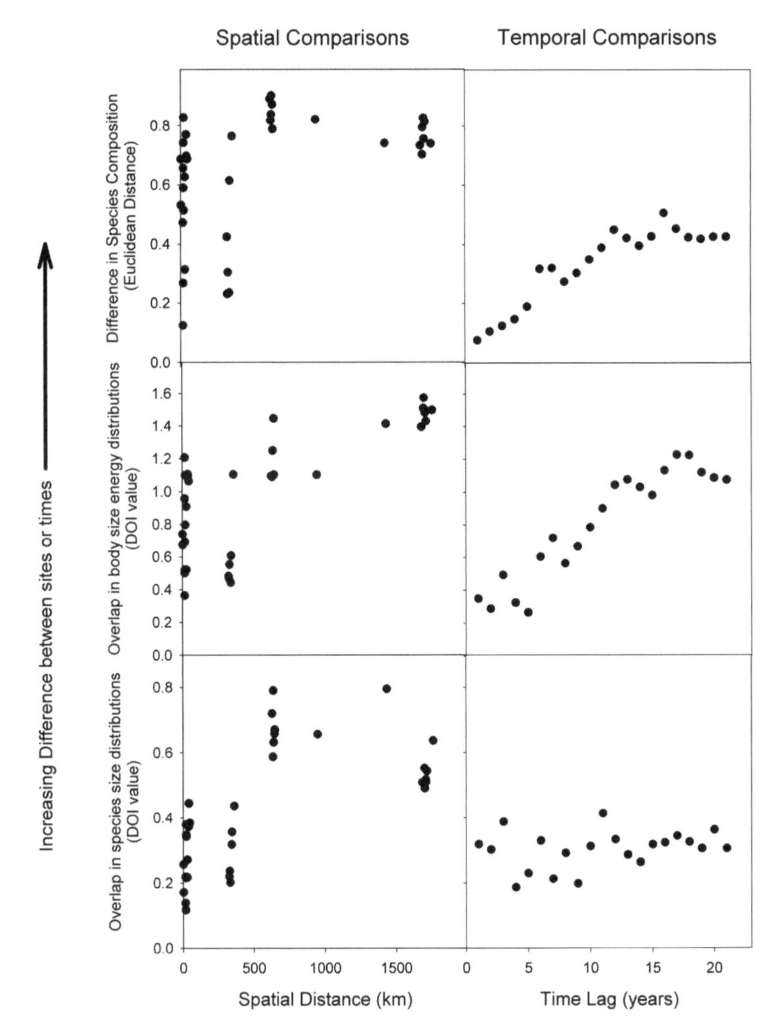


FIGURE 6.4. Temporal and spatial analyses of species composition, body size energy, and species size distributions. *Left*, Spatial comparisons showing the effect of spatial distance on the similarity between communities. Each point is a comparison between two communities from fig 6.3. *Right*, Temporal comparisons for a single site, the Portal Project. Each year was compared to the first year of data to determine how species composition and the shapes of the size distributions changed through time.

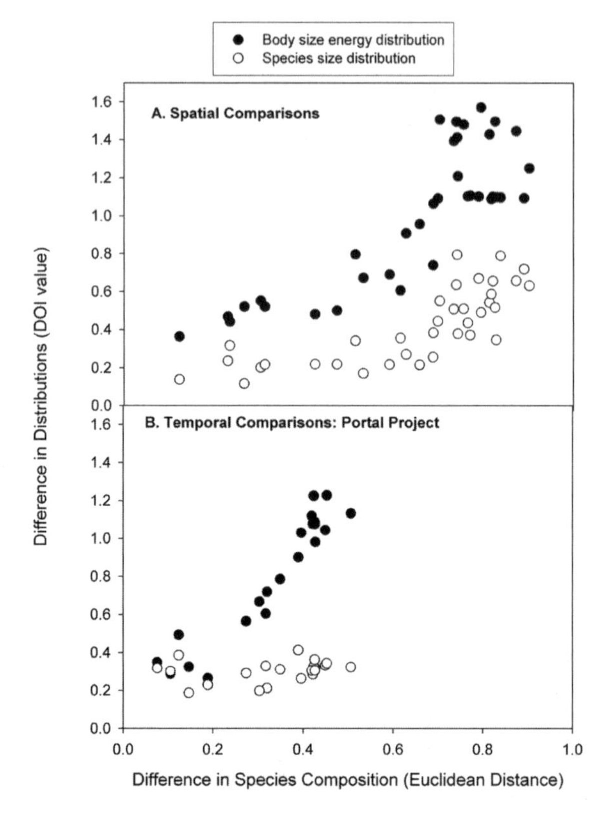


FIGURE 6.5. Relationship between similarity in species composition and similarity in body size and species size distributions. *Top*, Comparisons among the nine communities. *Bottom*, Temporal comparisons from the Portal Project.

complexity in the Southwest, sites located close together can be just as different from each other compositionally as sites located 1,000 km apart (fig 6.4). Similarity in species composition itself, rather than spatial distance, seems to provide a cleaner understanding of how similar the energy distribution will be between two sites (fig 6.5).

In contrast to the body size energy distributions, all sites had similar species size distributions (Kruskal-Wallis: all $p > 0.996$; Ernest 2005). These results are not surprising, since all sites also exhibited a species size distribution that was not significantly different from uniform (Ernest 2005). Overlap between sites in the shape of the species size distribution exhibited a complex relationship with the spatial distance between sites. Sites located less than 500 km apart appear to have more similar species size distributions than sites located over 500 km apart—however. this could be generated solely by the fact that sites within 500 km of one another were all located in the Chihuahuan Desert. As expected from these tests, differences in species composition had little effect upon the species size distribution; sites with large differences in species composition exhibited only modest differences in their body size distributions (fig. 6.5), and then only when sites were extremely different in species composition.

A similar picture emerges when the temporal dynamics of these distributions at the Portal Project are examined. At this site, species composition has changed directionally through time (fig. 6.4; Thibault et al. 2004). As expected from the spatial results, the body size energy distribution shows a strong directional shift in shape that is highly correlated with the shift in composition (fig 6.4). The species size distribution shows no directional shift either through time or in response to species composition changes. As the species composition at the site shifts, the body size energy distribution changes shape in response; however, the species size distribution again appears to be more resilient and exhibits less change in shape in response to the changes in species composition (fig 6.5).

These results suggest that different processes may be structuring the individual and species size distributions. Spatial and temporal analyses of the species size distribution suggest that it may be a relatively stable property of mammalian communities. Using this statistical approach, there is little or no detectable difference in the distribution of species masses either through time or across space. This suggests that at least up to some threshold of species compositional shift, the reorganization of species is relatively compensatory: when a species of a particular size drops out it is being replaced by a species of similar size. In contrast, the body size energy distribution was significantly different across sites and differences in the shape of the distribution show a tight relationship with differences in species composition, both across space and through time. Given the fact that the species composition at the Portal Project site has undergone strong directional shifts (fig. 6.4; Thibault et al. 2004; Ernest et al. 2008), including the extinction of a once-dominant species and the colonization and establishment of a new dominant species, it is interesting that the two distributions have responded to the compositional reorganization so differently. The more dramatic response of the body size energy distributions to changes in species composition could imply that while the overall species size distribution is a fairly stable property of mammalian communities, the importance of a particular body size for energy flux varies with changes in species composition and/or habitat structure.

Unfortunately, these analyses only demonstrate that the body size energy and species size distributions are differentially sensitive to shifts in species composition. They do not clarify whether ecosystem structure is the driving force behind the body size energy distribution. While these results potentially support the idea that the modes in energy use are potentially an idiosyncratic result of which species happen to be competitively dominant at a site, it could also be argued that shifts in habitat change how resources are available with respect to size and thus cause compositional shifts. Unfortunately there are very few replicates of a particular ecosystem type in the data used for this study and those replicates tend to be located in close spatial proximity. What is required to answer this question more definitively is a more extensive database of mammalian communities where multiple replicates of grasslands, shrublands, and forests can be analyzed to see if sites with more similar ecosystem structure also have more similar body size distributions.

## Conclusions

Despite their long histories, species and individual size distributions are rarely used in conjunction to study the same community. If anything, it is often assumed that both approaches are unnecessary; if there are many species of a particular size, there must also be many individuals packed into the same region of the body size axis. However, an abundance of species is not necessarily indicative of an abundance of individuals or a high level of resource use. Because these distributions are related but different, they provide very different perspectives on the role of size in structuring communities, as evidenced by the different responses of these distributions to shifting species composition, both across sites and through time at the same site (figs. 6.4 and 6.5). The highly conserved nature of species size distributions (Brown and Nicoletto 1991; Ernest 2005) suggests that this distribution may be a fundamental aspect of how mammals assemble, regardless of habitat type or species composition. In contrast, the sensitivity of the body size energy distribution to shifting species composition suggests that the shape of individual-based distributions may be more context specific. Whether the relevant context is the composition of the community or the structure of the ecosystem is still unclear. However, only by examining both distributions in concert will we gain fuller insight into both how species are assembled and the processes that determine which species or sizes dominate in ecological communities.

## Acknowledgments

I would like to thank Diane Ernest for extensive time management help and Ethan White for comments and discussion. Datasets provided by the Sevilleta LTER Data Bank and the Forest Science Data Bank were funded by the NSF LTER program (Sevilleta: BSR- 9411976 and DEB-0080529; H. J. Andrews: BSR-90–11663 and DEB-96–32921). Data were also provided by the NSF-funded Niwot Ridge LTER and the University of Colorado Mountain Research Station. The Portal Project (J. H. Brown, T. J. Valone, S. K. M. Ernest, Principal Investigators) was supported by the NSF LTREB Program. This work benefited from discussion/interaction with members of the IMPPS NSF-funded Research Coordination Network (DEB-0541625). This is IMPPS RCN publication no. 9.

## References

Allen, A. P., J. F. Gillooly, and J. H. Brown. 2005. "Linking the global carbon cycle to individual metabolism." *Functional Ecology* 19 (2): 202–213.

Allen, C. R., A. S. Garmestani, T. D. Havlicek, P. A. Marquet, G. D. Peterson, C. Restrepo, C. A. Stow, and B. E. Weeks. 2006. "Patterns in body mass distributions: Sifting among alternative hypotheses." *Ecology Letters* 9 (5): 630–643. doi: 10.1111/j.1461–0248.2006.00902.x.

Bakker, V. J., and D. A. Kelt. 2000. "Scale-dependent patterns in body size distributions of Neotropical mammals." *Ecology* 81 (12): 3530–3547.

Barnes, C., D. Maxwell, D. C. Reuman, and S. Jennings. 2010. "Global patterns in predator-prey size relationships reveal size dependency of trophic transfer efficiency." *Ecology* 91 (1): 222–232. doi: 10.1890/08–2061.1.

Belovsky, G. E. 1997. "Optimal foraging and community structure: The allometry of herbivore food selection and competition." *Evolutionary Ecology* 11 (6): 641–672.

Blaustein, A. R., and A. C. Risser. 1976. "Interspecific interactions between three sympatric species of kangaroo rats (*Dipodomys*)." *Animal Behaviour* 24 (2): 381–385.

Bowers, Michael A., and James H. Brown. 1982. "Body size and coexistence in desert rodents: Chance or community structure?" *Ecology* 63 (2): 391–400.

Brody, S. 1945. *Bioenergetics and growth*. New York: Hafner Press.

Brown, J. H. 1975. "Geographical ecology of desert rodents." In *Ecology and evolution of communities*, edited by M. L. Cody and J. M. Diamond, 315–341. Cambridge, MA: Harvard University Press.

Brown, J. H., P. A. Marquet, and M. L. Taper. 1993. "Evolution of body size: Consequences of an energetic definition of fitness." *American Naturalist* 142 (4): 573–584.

Brown, J. H., and P. F. Nicoletto. 1991. "Spatial scaling of species composition: Body masses of North American land mammals." *American Naturalist* 138 (6): 1478–1512.

Calder, W. A. 1984. *Size, function, and life history*. Cambridge, MA: Harvard University Press.

Collins, S. L. 2000. "Disturbance frequency and community stability in native tallgrass prairie." *American Naturalist* 155 (3): 311–325.

de Liocourt, F. 1898. "De l'amenagement des sapinières." *Bulletin trimestriel, Société forestière de Franche-Comté et Belfort*, July, 396–409.

Ernest, S. K. M. 2005. "Body size, energy use, and community structure of small mammals." *Ecology* 86 (6): 1407–1413.

Ernest, S. K. M., and J. H. Brown. 2001. "Homeostasis and compensation: The role of species and resources in ecosystem stability." *Ecology* 82 (8): 2118–2132.

Ernest, S. K. M., J. H. Brown, K. M. Thibault, E. P. White, and J. R. Goheen. 2008. "Zero sum, the niche, and metacommunities: Long-term dynamics of community assembly." *American Naturalist* 172 (6): E257–E269.

Ernest, S. K. M., B. J. Enquist, J. H. Brown, E. L. Charnov, J. F. Gillooly, V. Savage, E. P. White, F. A. Smith, E. A. Hadly, J. P. Haskell, S. K. Lyons, B. A. Maurer, K. J. Niklas, and B. Tiffney. 2003. "Thermodynamic and metabolic effects on the scaling of production and population energy use." *Ecology Letters* 6 (11): 990–995.

Gaston, K. J., and T. M. Blackburn. 1995. "The frequency distribution of bird body weights: Aquatic and terrestrial species." *Ibis* 137 (2): 237–240.

Gittleman, John L. 1985. "Carnivore body size: Ecological and taxonomic correlates." *Oecologia* 67 (4): 540–554. doi: 10.1007/bf00790026.

Griffiths, D. 1986. "Size-abundance relations in communities." *American Naturalist* 127 (2): 140–166.

Hemmingsen, A. M. 1960. "Energy metabolism as related to body size and respiratory surfaces, and its evolution." *Reports of the Steno Memorial Hospital and the Nordisk Insulinlaboratorium* 9:7–110.

Hersteinsson, P., and D. W. MacDonald. 1992. "Interspecific competition and the

geographical distribution of red and arctic foxes *Vulpes vulpes* and *Alopex lagopus*." *Oikos* 64 (3): 505–515.
Holling, C. S. 1992. "Cross-scale morphology, geometry, and dynamics of ecosystems." *Ecological Monographs* 62 (4): 447–502.
Hutchinson, G. E. 1959. "Homage to Santa Rosalia, or why are there so many kinds of animals?" *American Naturalist* 93 (870): 145–159.
Hutchinson, G. E., and R. H. MacArthur. 1959. "A theoretical ecological model of size distributions among species of animals." *American Naturalist* 93: 117–125.
Jablonski, D., and D. M. Raup. 1995. "Selectivity of end-Cretaceous marine bivalve extinctions." *Science* 268 (5209): 389–391.
Kerr, S. R., and L. M. Dickie. 2001. *The biomass spectrum*. New York: Columbia University Press.
Kleiber, M. 1932. "Body size and metabolism." *Hilgardia* 6:315–351.
MacArthur, R. H. 1958. "Population ecology of some warblers of northeastern coniferous forests." *Ecology* 39 (4): 599–619.
Manly, B. F. 1996. "Are there clumps in body-size distributions?" *Ecology* 77 (1): 81–86.
Marquet, P. A., and H. Cofre. 1999. "Large temporal and spatial scales in the structure of mammalian assemblages in South America: A macroecological approach." *Oikos* 85 (2): 299–309.
McNab, B. K. 2002. *The physiological ecology of vertebrates: A view from energetics*. Ithaca: Cornell University Press.
Morse, D. R., N. E. Stork, and J. H. Lawton. 1988. "Species number, species abundance and body length relationships of arboreal beetles in Bornean lowland rain-forest trees." *Ecological Entomology* 13 (1): 25–37.
Muller-Landau, H. C., R. S. Condit, K. E. Harms, C. O. Marks, S. C. Thomas, S. Bunyavejchewin, G. Chuyong, L. Co, S. Davies, R. Foster, S. Gunatilleke, N. Gunatilleke, T. Hart, S. P. Hubbell, A. Itoh, A. R. Kassim, D. Kenfack, J. V. LaFrankie, D. Lagunzad, H. S. Lee, E. Losos, J. R. Makana, T. Ohkubo, C. Samper, R. Sukumar, I. F. Sun, N. M. N. Supardi, S. Tan, D. Thomas, J. Thompson, R. Valencia, M. I. Vallejo, G. V. Munoz, T. Yamakura, J. K. Zimmerman, H. S. Dattaraja, S. Esufali, P. Hall, F. L. He, C. Hernandez, S. Kiratiprayoon, H. S. Suresh, C. Wills, and P. Ashton. 2006. "Comparing tropical forest tree size distributions with the predictions of metabolic ecology and equilibrium models." *Ecology Letters* 9 (5): 589–602. doi: 10.1111/J.1461-0248.2006.00915.X.
Niklas, K. J. 1994. *Plant allometry: The scaling of form and process*. Chicago: University of Chicago Press.
Owen-Smith, N. 1988. *Megaherbivores: The influence of very large body size on ecology*. Cambridge: Cambridge University Press.

Peters, R. H. 1983. *The ecological implications of body size.* Cambridge: Cambridge University Press.

Phillips, B. L., and R. Shine. 2004. "Adapting to an invasive species: Toxic cane toads induce morphological change in Australian snakes." *Proceedings of the National Academy of Sciences of the United States of America* 101 (49): 17150–17155. doi: 10.1073/Pnas.0406440101.

Ritchie, M. E. 1998. "Scale-dependent foraging and patch choice in fractal environments." *Evolutionary Ecology* 12 (3): 309–330.

Ritchie, M. E. , and H. Olff. 1999. "Spatial scaling laws yield a synthetic theory of biodiversity." *Nature* 400:557–560.

Rosenzweig, M. L. 1966. "Community structure in sympatric Carnivora." *Journal of Mammalogy* 47 (4): 602–612.

Rychlik, L., and R. Zwolak. 2006. "Interspecific aggression and behavioural dominance among four sympatric species of shrews." *Canadian Journal of Zoology* 84 (3) :434–448. doi: 10.1139/z06–017.

Savage, V. M., J. F. Gillooly, J. H. Brown, G. B. West, and E. L. Charnov. 2004. "Effects of body size and temperature on population growth." *American Naturalist* 163 (3): E429–E441.

Schmidt-Nielsen, K. 1984. *Scaling, why is animal size so important?* Cambridge: Cambridge University Press.

Schoener, T. W. 1968. "Sizes of feeding territories among birds." *Ecology* 49 (1): 123–141.

Schoener, T. W., and D. H. Janzen. 1968. "Notes on environmental determinants of tropical versus temperate insect size patterns." *American Naturalist* 102 (925): 207–224.

Sheldon, R. W., and T. R. Parsons. 1967. "A continuous size spectrum for particulate matter in the sea." *Journal of the Fisheries Research Board of Canada* 24 (5): 909–915.

Siemann, E., and J. H. Brown. 1999. "Gaps in mammalian body size distributions reexamined." *Ecology* 80 (8): 2788–2792.

Siemann, E., D. Tilman, and J. Haarstad. 1996. "Insect species diversity, abundance and body size relationships." *Nature* 380 (6576): 704–706.

———. 1999. "Abundance, diversity and body size: Patterns from a grassland arthropod community." *Journal of Animal Ecology* 68 (4): 824–835.

Smith, F. A., S. K. Lyons, S. K. M. Ernest, K. E. Jones, D. M. Kaufman, T. Dayan, P. A. Marquet, J. H. Brown, and J. P. Haskell. 2003. "Body mass of late Quaternary mammals." *Ecology* 84 (12): 3403–3403.

Sprules, W. G., and M. Munawar. 1986. "Plankton size spectra in relation to ecosystem productivity, size, and perturbation." *Canadian Journal of Fisheries and Aquatic Sciences* 43 (9): 1789–1794.

Thibault, K. M., E. P. White, and S. K. M. Ernest. 2004. "Temporal dynamics in the structure and composition of a desert rodent community." *Ecology* 85 (10): 2649–2655.

Thibault, K. M., E. P. White, A. H. Hurlbert, and S. K. M. Ernest. 2011. "Multimodality in the individual size distribution of bird communities." *Global Ecology and Biogeography* 20:145–153.

Thompson, D. W. 1917. *On growth and form.* Cambridge: Cambridge University Press.

Torres, L. E., and M. J. Vanni. 2007. "Stoichiometry of nutrient excretion by fish: Interspecific variation in a hypereutrophic lake." *Oikos* 116 (2): 259–270. doi: 10.1111/J.2006.0030–1299.15268.X.

Van Valen, L. M. 1973. "Body size and numbers of plants and animals." *Evolution* 27:27–35.

West, G. B., J. H. Brown, and B. J. Enquist. 1997. "A general model for the origin of allometric scaling laws in biology." *Science* 276 (5309): 122–126.

West, G. B., B. J. Enquist, and J. H. Brown. 2009. "A general quantitative theory of forest structure and dynamics." *Proceedings of the National Academy of Sciences of the United States of America* 106 (17): 7040–7045. doi: 10.1073/Pnas.0812294106.

White, E. P., S. K. M. Ernest, A. J. Kerkhoff, and B. J. Enquist. 2007. "Relationships between body size and abundance in ecology." *Trends in Ecology and Evolution* 22 (6): 323–330. doi: 10.1016/j.tree.2007.03.007.

White, E. P., S. K. M. Ernest, and K. M. Thibault. 2004. "Trade-offs in community properties through time in a desert rodent community." *American Naturalist* 164 (5): 670–676.

Wright, D. H.1983. "Species-energy theory: An extension of species-area theory." *Oikos* 41 (3): 496–506.

Yoda, K., T. Kira, H. Ogawa, and K. Hozumi. 1963. "Self-thinning in overcrowded pure stands under cultivated and natural conditions." *Journal of Biology, Osaka City University* 14:107–129.

CHAPTER SEVEN

# Processes Responsible for Patterns in Body Mass Distributions

Brian A. Maurer and Pablo A. Marquet

Collections of plants and animals that belong to the same taxon often exhibit striking variation in body masses, especially when the taxon being considered is a higher category such as an order or class. These large collections of species often exhibit striking statistical patterns in frequency distributions of logarithmically transformed body masses of species within the taxon (Brown 1981; Morse et al. 1985; Caughley 1987; Maurer and Brown 1988; Morse et al. 1988; Brown and Maurer 1989; Brown and Nicoletto 1991; Brown et al. 1993; Marquet and Cofre 1999; Roy et al. 2000; Smith et al. 2004). The variety of different taxa that exhibit these patterns, and the variation in the properties of frequency distributions among taxa, has led to a number of different explanations for why such patterns exist. Here we review these explanations and suggest that, rather than the proliferation of individual hypotheses, what is needed is an integration of different perspectives into a single, comprehensive framework.

Confusion has been generated in the past in part because researchers have sometimes compared hypotheses generated from data collected on different spatial scales or from different phylogenetic groups (Gaston and Blackburn 2000). The collection of species used to generate body size frequency distributions ranges from nearly complete collections of all species in a higher-level taxon (e.g., class) for the entire world to local collections of species found within a single landscape or ecosystem. Large-spatial-scale collections tend to consider monophyletic groups of species, while more localized collections examine paraphyletic species groups. There is no reason to suspect that a single process is responsible for all patterns. The approach we describe focuses on incorporating several suites of explanatory factors into a single framework, and on applying this framework to establish the relative contributions of different kinds of processes to the patterns in body size variation seen within particular collections of species.

## Hypotheses to Explain Patterns in Body Size Variation

As might be expected, the literature is replete with explanations for the existence of patterns seen in body mass distributions (Gardezi and da Silva 1999; Gaston and Blackburn 2000; Kozlowski and Gawelczyk 2002; Allen et al. 2006b). Many of these explanations refer to frequency distributions of log-transformed body masses, since untransformed body mass distributions are typically highly skewed "hollow curve" distributions. Researchers have identified four major patterns that need explanation (fig. 7.1). The first property is the maximum and minimum size of species in an assemblage. Depending on the taxonomic composition of the assemblage and the spatial scale at which it is examined, the range of these extremes can vary from a few orders of magnitude to over twenty (Brown et al. 2000). The second property is the modal, or most common, size. The modal size is almost always determined for log transformed body masses because of statistical problems with identifying modes in "hollow curve" distributions. The third property of log body mass distributions that has received attention is the shape. This focus on shape has primarily been on the degree of skewness of the frequency distribution. A "null" model for the frequency distribution of body masses of species is lognormal (McKinney 1990; McShea 1994; Maurer 1998a), since differences of body sizes among species could presumably be due to random differences in the rates of growth of different species, leading to random multiplicative differences among the adult (or maximum) sizes of different species. Finally, some authors have sought to explain the internal structure of the frequency distribution of body masses. This is commonly done by ranking species according to size to produce a body size *spectrum* for a species assemblage. Internal structure is then examined by seeking to identify discontinuities in the body mass spectrum

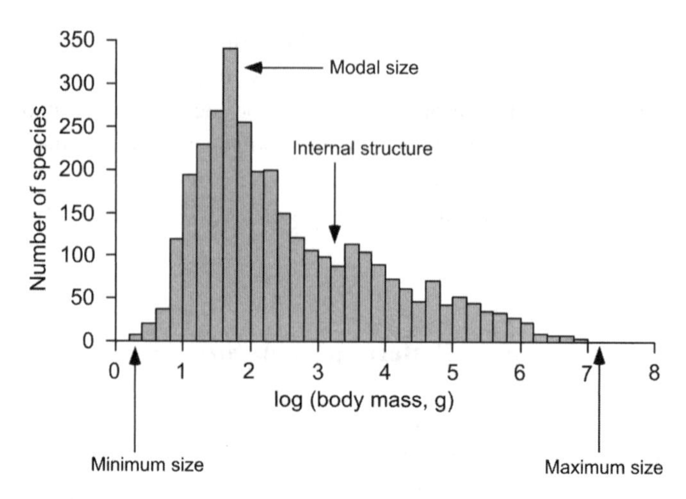


FIGURE 7.1. Properties of body mass distributions that various hypotheses attempt to explain. Data are of modern and late Pleistocene mammals of the world from Smith et al. (2004).

(Holling 1992; Siemann and Brown 1999). An alternative approach is to attempt to identify multiple modes within a log size frequency distribution (Manly 1996).

A number of explanations have been forwarded to explain these properties of body size distributions (table 7.1). Most explanations have focused on physiological, ecological or purely evolutionary processes, emphasizing one or the other of these types of factors. Recent reviews have discussed a number of explanations often favoring some above others (Gardezi and da Silva 1999; Gaston and Blackburn 2000; Kozlowski and Gawelczyk 2002; Allen et al. 2006). Indeed, since the physiological, ecological, and evolutionary processes involved in the generation of body size distributions are so complex, one could easily envision all explanations being right to a certain degree. Rather than emphasizing a single explanation, we propose to examine a single framework that incorporates all of the insights from these various models.

## A Synthetic Framework for Explaining Body Size Distributions

The fundamental concept that must underlie any framework for dealing with body size distributions is that they cannot exist without multiple

TABLE 7.1 **Explanations for Various Properties of Frequency Distributions of Sizes of Species within a Species Assemblage**

| Property | Explanation | Reference |
|---|---|---|
| Minimum | Constraints on oxygen supply (P) | Schmidt-Nielsen 1984 |
| | Constraints on pulsatile flow in supply networks (P) | West et al. 2002 |
| | Efficiency of ATP production (P) | Dobson and Headrick 1995 |
| | Competition (Ec) | Brown et al. 1978 |
| | Interactions among land mass area, energetic requirements, and space use (Ec) | Marquet and Taper 1998 |
| | Higher extinction rate for extreme-sized taxa (Ev) | Dial and Marzluff 1989 |
| Maximum | Efficiency of ATP production (P) | Dobson and Headrick 1995 |
| | Increased energetic costs of locomotion (P) | Garland 1983 |
| | Biomechanical and thermoregulatory constraints (P) | Schmidt-Nielsen 1984 |
| | Interactions among land mass area, energetic requirements, and space use (Ec) | Marquet and Taper 1998; Burness et al. 2001 |
| | Higher extinction rates for extreme-sized taxa (Ev) | Dial and Marzluff 1989 |
| Mode | Global fitness maximum (Ev) | Brown et al. 1993 |
| | Outcome of variation among species in conditions determining optimal size (Ev) | Kozlowski and Weiner 1997 |
| Shape | Trade-off between acquisition and allocation of energy for reproduction (P) | Brown et al. 1993 |
| | More ecological niches for small species (Ec) | Hutchinson and MacArthur 1959 |
| | Fractal structure of environment (Ec) | Morse et al. 1985 |
| | Cladogenetic diffusion from small ancestor (Ev) | Stanley 1973 |
| | Non-random diversification due to size-biased cladogenetic and anagentic change (Ev) | Maurer et al. 1992 |
| | Random evolutionary diffusion coupled with reflecting boundary at smallest size (Ev) | McShea 1994 |
| | Outcome of variation among species in conditions determining optimal size (Ev) | Kozlowski and Wiener 1997 |
| | Constraints on natural selection (Ev) | Maurer 1998b |
| Structure | Fractal structure of landscapes (Ec) | Holling 1992 |
| | Widespread species are of similar size (Ec) | Siemann and Brown 1999 |

*Note*: Explanations can be primarily physiological (P), ecological (Ec), or evolutionary (Ev).

pathways by which free energy obtained from light (or in some cases, inorganic compounds) is processed to do useful biological work (Odum 1994; Kozlowski and Gawelczyk 2002). Life is about acquiring, transforming, and allocating free energy to do useful biological work. Acquisition and allocation of energy, we claim, are the two most fundamental processes that underlie the size that a living entity attains. These two processes operate across all scales at which biological work is done in cells, organisms, populations, and species. Thus, size diversity reflects different solutions to the problem of obtaining and allocating free energy to do useful biological work. Energy capture occurs at quantum scales within biological systems, and this energy is then used to build and maintain vast networks of energy flow aggregated across a variety of fundamental spatial scales, the largest of which is the earth itself. Examination of these different spatial scales is essential to identifying important processes whose cumulative effects establish and maintain patterns in body size distributions across geographic space and within different taxa. Further, as we see in table 7.1, different explanations for body size distribution usually focus on one or a combination of these scales.

### *The Cellular Scale*

The most basic energy processing system known to biologists is the cell. Cells can vary in size over several orders of magnitude among different types of species, but for our purposes they can be considered to exist on a relatively small spatial scale ($\sim 10^{-6}$ m). Cells have a variety of physical structures and arrangements, but all share several energetic properties. First, all cells import relatively high-quality energy through some mode of energy capture (e.g., chemosynthesis, photosynthesis, passive or active diffusion). The mode of energy capture a cell uses depends on the immediate ecological circumstances in which the cell exists. Free-living cells must find reliable sources of energy in their environments, while cells contained within multicellular organisms often depend on diffusion processes to obtain energy. Second, energy captured by cells is used to maintain cellular structures (membranes, organelles, etc.) via complex biochemical pathways and enzyme cascades. Energy is also used to regulate these pathways via transcription and translation of information contained in nucleic acid complexes within cells. Finally, some energy captured by a cell is used for cellular replication. In free-living cells, this replication is directly controlled by the environment. In multicellular organisms, cellular reproduction is highly constrained by many factors that regulate the role that a particular cell plays in the overall functioning of the organism. Thus, the energetic functions of cells can be divided into two general classes of processes: those involved in energy acquisition and those involved in energy allocation. These two processes are not independent, because of all the energy that flows into a cell, some of it must be allocated back into acquisition if the cellular system is to persist. The relative importance of each of them, as well as the dominant cell size in free-living unicells, however, is regulated by resource availability (Grover 1991).

*The Organismal Scale*

A basic problem that must be solved by the cellular aggregates we call organisms is the dispensing of acquired energy to spatially remote cells within the organism. Simple organisms are built so that nearly all cells are close to nutrients accumulated by acquisition processes, so that molecular diffusion processes are sufficient to provide needed energy to each cell. The roughly Euclidean geometry of such organisms places a constraint on their size, since no cells can exist within the organism in locations that are too remote from nutrient reserves. In highly complex organisms with many diverse tissues located far from nutrient reserves, the spatial arrangement of tissues must take on a special form in order to allow nutrients to be delivered to individual cells. It turns out that the most efficient spatial organization of tissues to accomplish this is the construction of fractal-like networks of tubes that have a fixed minimal diameter equal roughly to the size of a cell (West et al. 1997, 1999b). This fundamental principle of organismal structure leads to a number of scaling relationships that exist among species across a wide range of sizes (Brown et al. 1997; West et al. 1997; Enquist et al. 1998; Enquist et al. 1999; West et al. 1999b, 1999a; Niklas and Enquist 2001; West et al. 2001; Enquist and Niklas 2002; Niklas and Enquist 2002a, 2002b; West et al. 2002).

The geometric constraints imposed by the fractal-like structure of multicellular organisms have profound consequences for the scaling of energy acquisition and allocation with body size. Acquisition of energy depends on the surface area of exchange surfaces across which energy-containing compounds diffuse to individual cells (West et al. 1999b; West et al. 2002). In the fractal-like networks through which these exchanges take place in organisms, this surface area must scale to the ¾ power of the volume containing the network (West et al. 1999b). Thus, energy acquisition in organisms increases proportional to organism size raised to the ¾ power. This scaling holds both within a species and among species because it reflects a fundamental design constraint for complex multicellular organisms. We examine the importance of this scaling later; here we emphasize that it reflects a fundamental physical constraint imposed upon organisms of all sizes and types as long as they adhere to the geometric structure required to deliver appropriate amounts of nutrients to individual cells. We also note here that the exchange surface problem also applies to the removal of cellular waste products (which also must diffuse short distances from cells) and to intercellular communication networks (e.g., nervous systems, endocrine systems, immune systems).

### *The Population Scale*

Populations are aggregates of individual organisms that share common genetic material and interact with one another to reproduce. Population processes are therefore the consequences of many events that happen to individual organisms as they collect, process, and partition energy across the space that the population occupies. The scaling of energetic processes with body size that characterizes the partitioning of acquired energy within organisms has important consequences for population processes. The rate of change of a population is determined by the rate of recruitment of new individuals and the rate of loss of existing individuals. In an isolated population, the rate of recruitment is determined by the rate of births. The rate of births is determined by the rate at which individual organisms are able to acquire and process energy. This rate is determined by how organisms in the population are able to partition energy into maintenance, growth, and reproduction. Thus, the scaling imposed upon organismal production by the geometric constraints inherent in their structure is reflected in the scaling of reproductive output with body size across a wide variety of species with very different life cycles (fig. 7.2).

This scaling of reproduction has important consequences for population dynamics. Even if mortality rate is independent of body size (Charnov 1993), the rate of change of a population must be constrained by body mass (Savage et al. 2004). There are two important consequences for population growth. In a rapidly growing population, $r_{max}$, the rate of

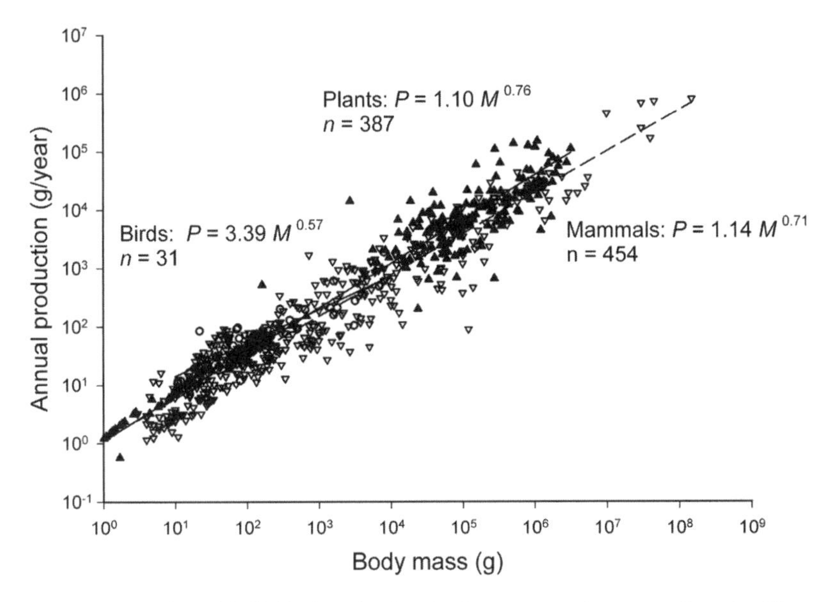


FIGURE 7.2. Scaling of individual production for three taxonomic groups of species. Note that the scaling exponents are similar for each taxon.

population change (assuming exponential growth) scales negatively with body size with an exponent of −¼ (Savage et al. 2004). That is, populations composed of large individuals cannot change as rapidly as populations of small individuals (Calder 1983; Peters 1983; Calder 1984). Second, for nongrowing populations, the steady state size a population can achieve must also scale negatively with body size, but with an exponent of −¾ (Savage et al. 2004). Such negative scaling of population density has been documented for a wide variety of taxa (Damuth 1981; Peters 1983; Brown and Maurer 1986; Damuth 1987, 1991; Brown 1995; Damuth 1998; Enquist et al. 1998, 1999; Cyr 2000).

Since population dynamics are constrained by body size, it follows that natural selection must also be fundamentally constrained by body size (Maurer 1998b, 2003). Models of natural selection are based on comparisons of the relative abilities of different heritable phenotypes to perpetuate themselves in populations exposed to a particular set of environmental conditions (Yodzis 1989; Roff 1992; Stearns 1992; Charnov 1993; Kozlowski 1996, 1999; Case 2000). Under this view, a population can be thought of as being an aggregate of populations of phenotypes, the dynamics of which are each constrained by size. The outcome of selection, then, is also constrained by size, even if size is not the primary character upon which natural selection works. Exactly how these constraints operate depends in large part on the environmental circumstances a population is found in. The scaling of organismal production implies a trade-off with increasing size between the ability to acquire and store energy, which increases with size, and the ability to transform resources into new offspring, which decreases with size (Brown et al. 1993; Maurer 1998b). Hence, natural selection in large organisms will be composed of phenotypic changes that enhance their ability to transform resources into new offspring. This is because the scaling of energy processing that results from the complex geometry of large organisms has forced them to the physical limits of energy acquisition and storage.

### *The Species Scale*

A species is a collection of populations that are recognized by some criterion of cohesion such as reproductive compatibility or common evolutionary history (Otte and Endler 1989; Baum 1998). The dynamic properties of a species must be determined by the combined effects of a vast number of individual events. In the preceding sections, we showed how the effects of body size on organismal and population processes were determined in large part by the physical constraints on energy flows that resulted from the complex geometry of organisms. It is not unreasonable to expect that the complex behavior of a species should also be influenced in a fundamental way by the size of the individuals that constitute it.

It has been proposed that species of different size have characteristic rates of speciation and extinction (McKinney 1990; Maurer et al. 1992; McShea 1994; Maurer 1998a; Kozlowski and Gawelczyk 2002). Extinction of an entire species is the consequence of a concerted pattern of extinctions of local populations across the geographic range of a species. It is unclear the degree to which extinction is influenced by the constraints on population processes that arise from organismal geometry. It is not entirely clear that local populations of large organisms are more likely to go extinct than those of small species (Pimm et al. 1988; Tracy and George 1992; Pimm et al. 1993). Consequently, it is not necessary that a large-sized species has a higher probability of going extinct than a small-sized species (Jablonski 1996). However, because the allometry of resource acquisition processes affects individuals' space requirements or home ranges (Brown and Maurer 1989), larger individuals (generally larger than about 100 g in mammals: Marquet and Taper 1998) require larger amounts of space to satisfy their energetic demands. Thus, given a particular amount of habitable area, a larger species will be able to pack fewer individuals into the area than a smaller species. All else being equal, this would increase the likelihood of extinction of the larger species. Marquet and Taper (1998) used this reasoning to explain the positive relationship between landmass and the size of the largest mammal inhabiting it (see also Burness et al. 2001), which implies that as landmass area decreases larger species tend to go extinct from extant biotas. Speciation is a more complex phenomenon than extinction, and although there may be some indirect effects of body size on speciation, through geographic range size, there is no evidence to suggest that there is a direct relationship between speciation and body size. Recent work points out that the link between body size and speciation is through the pervading effect of temperature and metabolism, which affect rates of DNA evolution and speciation (Gillooly et al. 2005; Allen et al. 2006a).

Despite the lack of evidence for the processes of speciation and extinction being affected directly by body size, there is evidence that the diversification of body sizes is not a purely random, epiphenomenal process (Maurer 1998a). Given the complexity of processes that occur over space and time to constitute the diversification process, it is probably overly simplistic to explain diversification solely as a consequence of a single phenotypic character (i.e., body size). Despite this, there must be a pervading effect of size, since it influences so many of the processes that affect diversification. A complementary approach would be to ask how size sets the boundaries within which diversification must occur (Maurer 1998b, 2003). From this perspective, it is clear that although there need not be a direct functional relationship between body size and diversification, nevertheless, body size can profoundly affect which types of organisms cannot exist within the genetic possibilities of an evolving lineage.

## Explanations for Body Size Distributions Revisited

In complex systems such as the earth's biota, there are numerous other criteria that can be used to identify hierarchical relationships among entities (MacMahon et al. 1978; Allen and Starr 1982; Eldredge 1985; Salthe 1985; Ahl and Allen 1996). In the search for explanations for patterns in

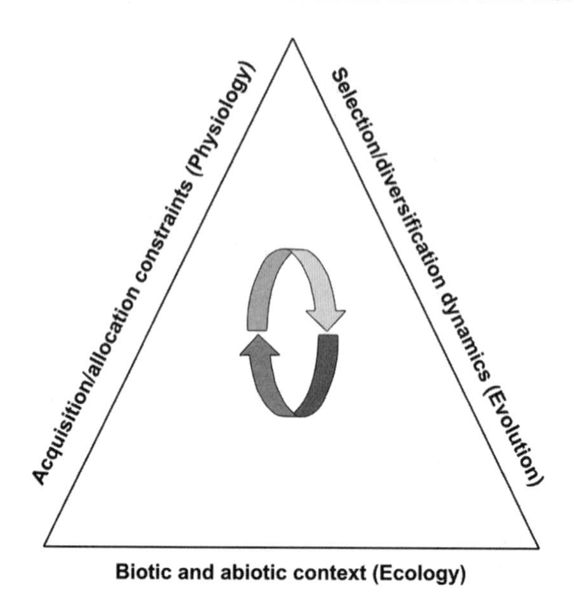


FIGURE 7.3. Schematic representation of three domains of explanation for body mass distributions. Any single explanation will incorporate each of these domains in some fashion or other, depending on the spatial, temporal, and taxonomic scale of the assemblage for which an explanation is required.

body mass distributions, the choice of a specific group of species and the region in space time in which they are examined have profound consequences for what explanations will make sense, and indeed, what aspects of body mass distributions must be explained (Gaston and Blackburn 2000). As shown in table 7.1, current explanations for body size distributions emphasize physiological, ecological, or evolutionary mechanisms, focusing on one or a combination of spatial scales as outlined in the previous section. However, because the physiological attributes of organisms are not independent of their ecological performance and evolutionary histories (i.e., they are not orthogonal axes), physiology, ecology, and evolution interact recursively in defining an explanatory domain for body mass patterns (fig. 7.3). And within this domain, the relative importance of the axes defining it (i.e., physiological, ecological, or evolutionary) will shift depending on the temporal, spatial, and taxonomic scales at which the patterns are analyzed. The geometric constraints operating upon acquisition and allocation of resources (physiology) interact with the distribution, abundance, and type of resource used, which are determined by the biotic (e.g., type and abundance of other consumers in the systems) and the abiotic environment wherein they are embedded (ecology), which provides a context for selection and diversification (evolution) of body size in time and space.

## The Way Ahead

Body size refers to the biomass at the individual level. A distribution of body sizes is just a collection of individual biomass values (usually lumped following some criteria). Its biological meaning will depend on the characteristics of the entities whose values are included in the analysis. A fundamental source of complication when dealing with a distribution of body sizes is that body size is correlated with many different physiological, ecological, and evolutionary processes and attributes (e.g., Peters 1983; Calder 1984; Schmidt-Nielsen 1984).

Hence, what is usually seen as a univariate distribution is in fact multivariate, meaning that it can be decomposed or teased apart in many different ways. For example, from a physiological perspective body size distributions can be analyzed with regard to the type of metabolism of the organism under analysis (e.g., endothermic vs. exothermic species), from an ecological perspective body size distributions can be constructed for different trophic groups or entire communities, and yet from an evolutionary perspective they can be constructed for different monophyletic lineages such as mammals. Interestingly, along each of the above-mentioned axes, individual biomass distributions can be further analyzed on how they change across time and space. The above implies that a body mass axis could be treated as an ordination axis, where variability along it will be the result of different explanatory variables each weighted by its relative contribution. Theoretically, only the distribution of the individual biomasses of all living entities that have ever existed on earth can be decomposed along all possible axes. Empirically, most distributions are assembled by fixing one axis, which limits further meaningful decompositions. Notwithstanding this difficulty, to further our understanding of body size distributions, it is necessary to recognize their multivariate character in order to characterize the way physiological, ecological, and evolutionary processes interact in determining its minimum, maximum, mode, and shape and how their relative importance varies as we tease them apart.

## Concluding Remarks

In general terms, it is safe to say that the size of a biological system affects most processes that take place within it and in interaction with its biotic and abiotic environment by regulating fluxes of energy and materials (Brown et al. 2004; Marquet et al. 2005; West and Brown 2005). However, this is not a one-way process, for the size of a system is also affected by the internal and external process it regulates, in a process of mutual codetermination or circular causality (Hutchinson 1948; Maturana and Varela 1987; Odling-Smee et al. 2003; Marquet et al. 2004; Marquet et al. 2005). The logic behind this statement is rooted in the emergent character of size. By emergent we signify that the size of any biological system cannot be explained or predicted by observation of its components (Salt 1979), but emerges as the result of the operation of many entangled processes including how components interact among themselves and with the external environment. For the sake of generality we can group these processes in three major categories: (1) external processes and states, related to the biotic and abiotic environment wherein the target biological entity unfolds (we use the phrase "biological entity" to designate any biological self-replicating system that possesses well-defined boundaries, from a subcellular organelle up to the biosphere), (2) internal processes and states, or those built-in processes necessary to maintain the biological entity's structure, function, and fitness, and finally, (3) historical processes that describe the trajectory of the two-way interaction between the biological entity and its environment through time (the history of the interaction between internal and external processes and states). These three kinds of processes define the phase space wherein the system drifts (see fig. 7.3) or, more precisely, where we as observers make it understandable (i.e., they define an explanatory domain). In principle, any attribute of a biological entity (such as the number of components it possesses or the number of species in the context of an ecosystem) can be understood, at least in principle, using any of these three major axes as a departure point (Marquet et al. 2004), and will occupy a position in this explanatory domain reflecting the relative contribution of each of these axes to the pattern or attribute under investigation.

Biodiversity is pretty much about diversity in form, function, and organization. That part of this diversity can be mapped to one variable (e.g., individual biomass) is a blessing as well as a challenge not only in terms of statistical approaches, but, foremost, also in terms of data collection as well as theoretical frameworks for understanding. Further advances in our understanding of individual biomass distributions require exhaustive sampling of entire animal and plant assemblages and even simple ecosystems. Individual biomass distributions at this level would allow us meaningful decomposition and understanding of how ecological systems work.

## References

Ahl, V., and T. F. H. Allen. 1996. *Hierarchy theory : A vision, vocabulary, and epistemology.* New York: Columbia University Press.

Allen, A. P., J. F. Gillooly, V. M. Savages, and J. H. Brown. 2006. "Kinetic effects of temperature on rates of genetic divergence and speciation." *Proceedings of the National Academy of Sciences of the United States of America* 103 (24): 9130–9135.

Allen, C. R., A. S. Garmestani, T. D. Havlicek, P. A. Marquet, G. D. Peterson, C. Restrepo, C. A. Stow, and B. E. Weeks. 2006. "Patterns in body mass distributions: Sifting among alternative hypotheses." *Ecology Letters* 9 (5): 630–643. doi: 10.1111/j.1461-0248.2006.00902.x.

Allen, T. F. H., and T. B. Starr. 1982. *Hierarchy : Perspectives for ecological complexity.* Chicago: University of Chicago Press.

Baum, D. A. 1998. "Individuality and the existence of species through time." *Systematic Biology* 47 (4): 641–653.

Brown, J. H. 1981. "Two decades of homage to Santa Rosalia: Toward a general theory of diversity." *American Zoologist* 21:877–888.

———. 1995. *Macroecology.* Chicago: University of Chicago Press.

Brown, J. H., W. A. Calder, and A. Kodricbrown. 1978. "Correlates and consequences of body size in nectar-feeding birds." *American Zoologist* 18 (4): 687–700.

Brown, J. H., B. J. Enquist, and G. B. West. 1997. "Allometric scaling laws in biology: Response." *Science* 278 (5337): 372–373.

Brown, J. H., J. F. Gillooly, A. P. Allen, V. M. Savage, and G. B. West. 2004. "Toward a metabolic theory of ecology." *Ecology* 85 (7): 1771–1789.

Brown, J. H., P. A. Marquet, and M. L. Taper. 1993. "Evolution of body size: Consequences of an energetic definition of fitness." *American Naturalist* 142 (4): 573–584.

Brown, J. H., and B. A. Maurer. 1986. "Body size, ecological dominance and Cope's rule." *Nature* 324 (6094): 248–250.

——. 1989. "Macroecology: The division of food and space among species on continents." *Science* 243 (4895): 1145–1150.

Brown, J. H., and P. F. Nicoletto. 1991. "Spatial scaling of species composition: Body masses of North-American land mammals." *American Naturalist* 138 (6): 1478–1512.

Brown, J. H., G. B. West, and Santa Fe Institute (Santa Fe, N.M.). 2000. *Scaling in biology*. Oxford: Oxford University Press.

Burness, G. P., J. Diamond, and T. Flannery. 2001. "Dinosaurs, dragons, and dwarfs: The evolution of maximal body size." *Proceedings of the National Academy of Sciences of the United States of America* 98 (25): 14518–14523.

Calder, W. A. 1983. "Ecological scaling: Mammals and birds." *Annual Review of Ecology and Systematics* 14:213–230.

——. 1984. *Size, function, and life history*. Cambridge, MA: Harvard University Press.

Case, Ted J. 2000. *An illustrated guide to theoretical ecology*. New York: Oxford University Press.

Caughley, G. 1987. "The distribution of eutherian body weights." *Oecologia* 74:319–320.

Charnov, E. L. 1993. *Life history invariants : Some explorations of symmetry in evolutionary ecology*. Oxford Series in Ecology and Evolution. Oxford: Oxford University Press.

Cyr, H. 2000. "Individual energy use and the allometry of population density." In *Scaling in biology*, edited by J. H. Brown and G. B. West, 267–295. New York: Oxford University Press.

Damuth, J. D. 1981. "Population density and body size in mammals." *Nature* 290:699–700.

——. 1987. "Interspecific allometry of population density in mammals and other animals: The independence of body mass and population energy-use." *Biological Journal of the Linnean Society* 31:193–246.

——. 1991. "Of size and abundance." *Nature* 351:268–269.

——. 1998. "Common rules for animals and plants." *Nature* 395:115–116.

Dial, K. P., and J. M. Marzluff. 1989. "Nonrandom diversification within taxonomic assemblages." *Systematic Zoology* 38 (1): 26–37.

Dobson, G. P., and J. P. Headrick. 1995. "Bioenergetic scaling: Metabolic design and body-size constraints in mammals." *Proceedings of the National Academy of Sciences of the United States of America* 92 (16): 7317–7321.

Eldredge, Niles. 1985. *Unfinished synthesis : Biological hierarchies and modern evolutionary thought*. New York: Oxford University Press.

Enquist, B. J., J. H. Brown, and Geoffrey B. West. 1998. "Allometric scaling of plant energetics and population density." *Nature* 395:163–165.

——. 1999. "Plant energetics and population density." *Nature* 398:573.

Enquist, B. J., and K. J. Niklas. 2002. "Global allocation rules for patterns of biomass partitioning in seed plants." *Science* 295 (5559): 1517–1520.

Enquist, B. J., G. B. West, E. L. Charnov, and J. H. Brown. 1999. "Allometric scaling of production and life-history variation in vascular plants." *Nature* 401 (6756): 907–911.

Gardezi, T., and J. da Silva. 1999. "Diversity in relation to body size in mammals: A comparative study." *American Naturalist* 153 (1): 110–123.

Garland, T. 1983. "Scaling the ecological cost of transport to body mass in terrestrial mammals." *American Naturalist* 121 (4): 571–587.

Gaston, K. J., and T. M. Blackburn. 2000. *Pattern and process in macroecology.* Oxford: Blackwell Science.

Gillooly, J. F., A. P. Allen, G. B. West, and J. H. Brown. 2005. "The rate of DNA evolution: Effects of body size and temperature on the molecular clock." *Proceedings of the National Academy of Sciences of the United States of America* 102 (1): 140–145.

Grover, J. P. 1991. "Resource competition in a variable environment: Phytoplankton growing according to the variable-internal-stores model." *American Naturalist* 138:811–835.

Holling, C. S. 1992. "Cross-scale morphology, geometry, and dynamics of ecosystems." *Ecological Monographs* 62 (4): 447–502.

Hutchinson, G. E. 1948. "Circular causal systems in ecology." *Annals of the New York Academy of Sciences* 50:221–246.

Hutchinson, G. E., and R. H. MacArthur. 1959. "A theoretical ecological model of size distributions among species." *American Naturalist* 93:117–125.

Jablonski, D. 1996. "Body size and macroevolution." In *Evolutionary paleobiology*, edited by D. Jablonski, D. H. Erwin and J. H. Lipps, 256–289. Chicago: University of Chicago Press.

Kozlowski, J. 1996. "Optimal allocation of resources explains interspecific life-history patterns in animals with indeterminate growth." *Proceedings of the Royal Society B: Biological Sciences* 263 (1370): 559–566.

——. 1999. "Adaptation: A life history perspective." *Oikos* 86 (2): 185–194.

Kozlowski, J., and A. T. Gawelczyk. 2002. "Why are species' body size distributions usually skewed to the right?" *Functional Ecology* 16 (4): 419–432.

Kozlowski, J., and J. Weiner. 1997. "Interspecific allometries are by-products of body size optimization." *American Naturalist* 149 (2): 352–380.

MacMahon, J. A., D. L. Phillips, J. V. Robinson, and D. J. Schimpf. 1978. "Levels of biological organization: Organism-centered approach." *Bioscience* 28 (11): 700–704.

Manly, B. F. J. 1996. "Are there clumps in body-size distributions?" *Ecology* 77 (1): 81–86.

Marquet, P. A., and H. Cofre. 1999. "Large temporal and spatial scales in the structure of mammalian assemblages in South America: A macroecological approach." *Oikos* 85 (2): 299–309.

Marquet, P. A., M. Fernández, S. A. Navarrete, and C. Valdivinos. 2004. "Diversity emerging: Towards a deconstruction of biodiversity patterns." In *Frontiers of biogeography: New directions in the geography of nature*, edited by M. V. Lomolino and L. R. Heaney, 436. Sunderland, MA: Sinauer Associates.

Marquet, P. A., R. A. Quinones, S. Abades, F. Labra, M. Tognelli, M. Arim, and M. Rivadeneira. 2005. "Scaling and power-laws in ecological systems." *Journal of Experimental Biology* 208 (9): 1749–1769.

Marquet, P. A., and M. L. Taper. 1998. "On size and area: Patterns of mammalian body size extremes across landmasses." *Evolutionary Ecology* 12 (2): 127–139.

Maturana, H. R., and F. J. Varela. 1987. *The tree of knowledge: The biological roots of human understanding*. Boston: Shambhala.

Maurer, B. A. 1998a. "The evolution of body size in birds. I. Evidence for non-random diversification." *Evolutionary Ecology* 12:925–934.

———. 1998b. "The evolution of body size in birds. II. The role of reproductive power." *Evolutionary Ecology* 12:935–944.

———. 2003. "Adaptive diversification of body size: The roles of physical constraint, energetics, and natural selection." In *Macroecology: Causes and consequences*, edited by T. M. Blackburn and K. J. Gaston, 174–191. Oxford: Blackwell.

Maurer, B. A., and J. H. Brown. 1988. "Distribution of energy use and biomass among species of North American terrestrial birds." *Ecology* 69 (6): 1923–1932.

Maurer, B. A., J. H. Brown, and R. D. Rusler. 1992. "The micro and macro in body size evolution." *Evolution* 46 (4): 939–953.

McKinney, M. L. 1990. "Trends in body size evolution." In *Evolutionary trends*, edited by K. J. McNamara, 75–118. Tucson: University of Arizona Press.

McShea, D. W. 1994. "Mechanisms of large-scale evolutionary trends." *Evolution* 48 (6): 1747–1763.

Morse, D. R., J. H. Lawton, M. M. Dodson, and M. H. Williamson. 1985. "Fractal dimension of vegetation and the distribution of arthropod body lengths." *Nature* 314 (6013): 731–733.

Morse, D. R., N. E. Stork, and J. H. Lawton. 1988. "Species number, species abundance and body length relationships of arboreal beetles in Bornean lowland rain forest trees." *Ecological Entomology* 13 (1): 25–27.

Niklas, K. J., and B. J. Enquist. 2001. "Invariant scaling relationships for interspecific plant biomass production rates and body size." *Proceedings of the National Academy of Sciences of the United States of America* 98 (5): 2922–2927.

——. 2002a. "Canonical rules for plant organ biomass partitioning and annual allocation." *American Journal of Botany* 89 (5): 812–819.

——. 2002b. "On the vegetative biomass partitioning of seed plant leaves, stems, and roots." *American Naturalist* 159 (5): 482–497.

Odling-Smee, F. J., K. N. Laland and M. W. Feldman. 2003. *Niche construction: The neglected process in evolution*. Princeton: Princeton University Press.

Odum, H. T. 1994. *Ecological and general systems : An introduction to systems ecology*. Rev. ed. Niwot: University Press of Colorado.

Otte, D., and J. A. Endler. 1989. *Speciation and its consequences*. Sunderland, MA: Sinauer Associates.

Peters, R. H.. 1983. *The ecological implications of body size*. Edited by E. Beck, H. J. B. Birks, and E. F. Connor. 1st ed. Cambridge Studies in Ecology. Cambridge: Press Syndicate of the University of Cambridge.

Pimm, S. L., J. Diamond, T. M. Reed, G. J. Russell, and J. Verner. 1993. "Times to extinction for small populations of large birds." *Proceedings of the National Academy of Sciences of the United States of America* 90 (22): 10871–10875.

Pimm, S. L., H. L. Jones, and J. Diamond. 1988. "On the risk of extinction." *American Naturalist* 132 (6): 757–785.

Roff, D. A. 1992. *The evolution of life histories: Theory and analysis*. New York: Chapman and Hall.

Roy, K., D. Jablonski, and K. K. Martien. 2000. "Invariant size- frequency distributions along a latitudinal gradient in marine bivalves." *Proceedings of the National Academy of Sciences of the United States of America* 97 (24): 13150–13155.

Salt, G. W. 1979. "A comment on the use of the term emergent properties." *American Naturalist* 113:145–149.

Salthe, Stanley N. 1985. *Evolving hierarchical systems: Their structure and representation*. New York: Columbia University Press.

Savage, V. M., J. F. Gillooly, J. H. Brown, G. B. West, and E. L. Charnov. 2004. "Effects of body size and temperature on population growth." *American Naturalist* 163 (3): E429–E441.

Schmidt-Nielsen, K. 1984. *Scaling, why is animal size so important?* Cambridge: Cambridge University Press.

Siemann, E., and J. H. Brown. 1999. "Gaps in mammalian body size distributions reexamined." *Ecology* 80 (8): 2788–2792.

Smith, F. A., J. H. Brown, J. P. Haskell, S. K. Lyons, J. Alroy, E. L. Charnov, T. Dayan, B. J. Enquist, S. K. M. Ernest, E. A. Hadly, K. E. Jones, D. M. Kaufman, P. A. Marquet, B. A. Maurer, K. J. Niklas, W. P. Porter, B. Tiffney, and M. R. Willig. 2004. "Similarity of mammalian body size across the taxonomic hierarchy and across space and time." *American Naturalist* 163 (5): 672–691.

Stanley, S. M. 1973. "An explanation for Cope's rule." *Evolution* 27:1–26.

Stearns, S. C. 1992. *The evolution of life histories.* Oxford: Oxford University Press.

Tracy, C. R., and T. L. George. 1992. "On the determinants of extinction." *American Naturalist* 139 (1): 102–122.

West, G. B., and J. H. Brown. 2005. "The origin of allometric scaling laws in biology from genomes to ecosystems: Towards a quantitative unifying theory of biological structure and organization." *Journal of Experimental Biology* 208 (9): 1575–1592.

West, G. B., J. H. Brown, and B. J. Enquist. 1997. "A general model for the origin of allometric scaling laws in biology." *Science* 276 (5309): 122–126.

——. 1999a. "A general model for the structure and allometry of plant vascular systems." *Nature* 400 (6745): 664–667.

——. 1999b. "The fourth dimension of life: Fractal geometry and allometric scaling of organisms." *Science* 284:1677–1679.

——. 2001. "A general model for ontogenetic growth." *Nature* 413 (6856): 628–631.

West, G. B., W. H. Woodruff, and J. H. Brown. 2002. "Allometric scaling of metabolic rate from molecules and mitochondria to cells and mammals." *Proceedings of the National Academy of Sciences of the United States of America* 99:2473–2478.

Yodzis, Peter. 1989. *Introduction to theoretical ecology.* New York: Harper and Row.

CHAPTER EIGHT

# The Influence of Flight on Patterns of Body Size Diversity and Heritability

Felisa A. Smith, S. Kathleen Lyons, Kate E. Jones,
Brian A. Maurer, and James H. Brown

The ability to fly confers sufficient selective benefits that over evolutionary time three different vertebrate groups have independently evolved powered flight. Of these three—the Pterosauria (pterosaurs), Chiroptera (bats), and Aves (birds)—the latter have unquestionably been the most successful, diversifying into an enormous range and number of species. Current taxonomy places the number of recognized bird species at over 10,000 (Clements 2007). Although Chiroptera also evolved powered flight and are relatively speciose (~1,100 recognized species; Wilson and Reeder 1993), they occupy a much narrower range of the body mass spectrum and are not nearly as ecologically diverse (Fenton 2001; Maurer et al. 2004). Most Chiroptera, for example, are either frugivorous (e.g., Megachiroptera), or insectivorous, and all are nocturnal. In contrast, birds occupy virtually all trophic levels from scavengers (e.g., buzzards, vultures) to top predators (e.g., falcons, hawks, and their allies) and have evolved enormous diversity in both size and shape.

The development of powered flight brought with it a common set of physiological, morphological, and ecological constraints (Calder 1984; Alexander 2003; Pennycuick 2008). Because flight evolved independently, taxa "solved" these constraints in different ways, resulting in significant differences in life history strategies among taxa. All bats, for example, have extremely low fecundity, with the majority of species having only one offspring per year (Jones and MacLarnon 2001). Moreover, they have much increased longevity relative to their body size (Jones and

MacLarnon 2001). Birds are much more variable in life history characteristics, although they generally tend to be more fecund and shorter lived than bats. For example, although an albatross may lay an egg once every two years and has a life span exceeding 50 years, partridges average 15–19 eggs per clutch, and under particularly good environmental conditions zebra finches reportedly raise more than 12 broods in a single year (Zann 1996). The variation in life history traits among bird species reflects fundamental differences in how energy is acquired and allocated to the essential activities of maintenance, growth, and reproduction.

One inescapable constraint facing all volant animals is that of body size (Calder 1984; Brown 1995; Maurer 1999; Alexander 2003). Interestingly, minimum body size is about the same in both volant and nonvolant mammals and in birds. For example, the smallest extant bird is the bee hummingbird (*Mellisuga helenae*) of Cuba, at about 1.6 g, nearly identical in mass to the bumblebee bat (*Craseonycteris thonglongyai*) of Thailand and the smallest nonvolant mammal, the pygmy white-toothed shrew (*Suncus etruscus*) of Asia and Europe (Smith et al. 2003). These similarities suggest that common constraints, which may be unrelated to the ability to fly, act to limit the minimum body size of vertebrate animals.

In contrast, the ability to fly clearly limits upper body size because of aerodynamic and biomechanical considerations (Calder 1984; Marden 1994; Templin 2000; Alexander 2003; Pennycuick 2008). The largest volant mammal, the golden-capped fruit bat (*Acerodon jubatus*, at 1.2 kg; Heaney and Heideman 1987) is more than an order of magnitude smaller than the largest known bird. Such large differences in the maximum size of birds and flying mammals probably reflects better adaptations to flight on the part of birds. However, mass is not the only limiting criterion; the size and structure of the wing are also critical (McGowan and Dyke 2007). The term "wing loading" (the mass of the animal divided by the area of the wing) is often used as an index of flight performance, with a ratio of 25 kg/m$^2$ considered the upper limit for flight (Meunier 1951). Larger animals must generate more power to obtain the same lift, because of the allometric scaling of muscle mass specific power; at some point, this exceeds lift production capacity (Marden 1994). Moreover, large size generally translates into high weight and drag forces as well as decreased thrust efficiency. Thus, the largest flying animals are not predominantly powered fliers (Pennycuick 2008; Sato et al. 2009). Examination of morphology and wing-loading ratios for extinct Pterosaurs, for example, suggest they probably largely relied on gliding (Templin 2000; Sato et al. 2009). This allowed evolution of enormous size with wingspans of 12 m or more. It is probable that Pterosaurs were heavily dependent on thermal currents and prevailing winds for locomotion. The largest volant bird, the giant teratorn of the late Miocene, had wingspans up to 8 m and probably also relied heavily on soaring, with only limited use of powered flight (Campbell and Tonni 1980; Marden 1994; Vizcaíno and Fariña 1999). At approximately 65–100 kg, the giant teratorn was 2–5 times more massive than the heaviest extant flying bird, the African kori bustard (*Ardeotis kori*). Both birds and pterosaurs share a number of flight adaptations that are not found in bats. These include strong but hollow bones, which serve to minimize mass, keeled sterna where flight muscles attach, and feathers/wing fibers.

Here, we address the influence of flight on patterns of body size diversification and evolution for birds and bats. Clearly, the ability to fly has required body plan modification and design elements not found in nonvolant mammals. Such an "adaptive syndrome" (Niklas 2000; Price 2003) may well result in common ecological and evolutionary constraints among nonrelated vertebrates that share the innovation. Indeed, similarities in the body mass distributions among volant species of distantly related vertebrate taxa have been attributed to the functional constraints of flight (Maurer et al. 2004), and so we might expect congruence in evolutionary patterns. For mammals, we find consistent macroecological patterns in body size that persist across both geographic space and evolutionary time (Smith et al. 2004). The body size "niche" of orders is remarkably consistent across the continents, for example, despite virtually no taxonomic overlap (Smith et al. 2004; F. A. Smith, unpublished data). This suggests a strong role of ecology in shaping body size distributions. However, we also find strong conservation of body size within the taxonomic hierarchy: estimates of broad-sense "heritability" for species with orders, and orders within families, exceed 0.9 (Smith et al. 2004). Such congruence in body size among sibling species implies strong selective pressures, which could be phylogenetic or ecological in nature. Interestingly, while heritability estimates are invariant across the majority of the size spectrum, there is *no* significant relationship between the body size of sister species for mammals under ~18 g (Smith et al. 2004). The lack of heritability at very small body sizes may well reflect increasingly tight allometric constraints on life history and ecological parameters, which reduce the ability to adapt to novel ecological conditions in ways other than by modifying size (Smith et al. 2004). Hence, both ecology and phylogeny appear to be important determinants of the patterns of body size diversification in mammals.

To date, similar studies are lacking for other vertebrates, including birds. Thus, the universality of such patterns is unclear. Do the remarkably consistent macroecological patterns of body size reported for mammals reflect fundamental ecological processes acting on vertebrates, or are they the result of the highly deterministic growth pattern of mammals? And how might a major innovation such as flight influence and/or constrain such patterns of heritability and vertebrate body size diversification? Flight was not explicitly examined in earlier work on the macroecological patterns of body size diversification. Do the patterns of heritability and diversification for bats resemble more closely those of other mammals, as might be expected (e.g., Smith et al. 2004), or do they resemble those of birds, with whom they share a major evolutionary innovation?

## Data and Analysis

Our mammal analyses utilized an updated version of a comprehensive database of body mass and taxonomy originally compiled by the NCEAS Working Group "Body Size in Ecology and Paleoecology" (Smith et al. 2003). The most recent version (MOM v3.6) assembles standardized data on body mass and taxonomy for all late Quaternary mammals of the world using a variety of literature sources, mammalian species accounts, measurements of museum specimens, and unpublished field notes; 5,755 species are represented, including insular, aquatic, and extinct forms. Rather than using a generic mean to represent species missing reliable mass estimates, they were excluded from any analyses. A single body mass estimate was derived for each species by averaging male and female body mass in grams across the geographic distribution. The body mass range represented in the database varies from 0.25 (1.8 g) for the shrew *Suncus etruscus* to 8.279 log units (190,000,000 g) for the blue whale, *Balaenoptera musculus*. For the present analysis, data were restricted to terrestrial mammals on the four major continents (North and South America, Eurasia, and Africa). All data were log transformed prior to analysis (see table 8.1 for summary statistics).

TABLE 8.1 **Sample Size, Moments, and Other Descriptive Statistics**

| | N | Median | Mode | Mean | SD | Range | Skew | Kurtosis |
|---|---|---|---|---|---|---|---|---|
| Nonvolant mammals (North America, South America, Africa, Eurasia) | 2,769 | 2.298 | 1.602 | 2.662 | 1.36 | 6.757 | 0.823 | −0.101 |
| Bats | 686 | 1.099 | 0.724 | 1.165 | 0.416 | 2.710 | 1.143 | 2.162 |
| Birds | 5,877 | 1.568 | | 1.716 | 0.696 | 3.80 | | |

Bird mass was obtained from the CRC handbook (Dunning 1993). As described above, data were averaged across the geographic range and then by gender to arrive at an estimate for each species. Because we were interested in the influence of flight on patterns of body size, all nonvolant birds were removed from our analyses. These primarily consisted of ratites, penguins, and a number of indigenous birds from New Zealand, New Guinea, and other island habitats. All body masses estimates (N = 6,016) were log transformed prior to analysis. Maurer (chap. 3 above) analyzed the same data but did so using all species of birds, including nonflying forms.

Phylogenetic relationships among birds and among mammals were approximated by using existing taxonomies. This was because no fully resolved species-level phylogeny is presently available for either class. Taxonomic affiliations followed Wilson and Reeder (1993) for mammals, and Clements (1991) for birds. ANOVAs were also conducted on Sibley and Ahlquist (1990) to examine the influence of a divergent (and somewhat controversial) taxonomy based largely on genetic data. Note that a number of assumptions were implicit in our use of both bird and mammal taxonomies as proxies for phylogeny. First, branch length was assumed to be invariant. That is, we assumed equivalent evolutionary distances between all sibling species, genera, and families and, further, that nodes were represented by hard polytomies. Second, we assumed evolutionary ages of genera, families, or orders were unrelated to the pattern of body size diversification. Both of these assumptions probably resulted in *underestimates* of heritability, because older species within a taxon could diverge more because of longer opportunities to do so. Third, we assumed that the *size* of a taxon was not related to its evolutionary age. Earlier work suggested that these assumptions were reasonably robust (Smith et al. 2004).

### *Statistical Analyses*

To address the issue of whether bats are in essence flying rats or furry birds, our analysis employed a variety of statistical techniques including "sib-sib" regression, phylogenetic autocorrelation (Moran's *I*), and nested analysis of variance. Each method yields different insights into the relationship between body size and the taxonomic hierarchy and—since taxonomy reflects phylogeny—how phylogenetic constraints interact with other processes to influence evolutionary diversification.

### *Sib-Sib Regressions*

We examined the degree of similarity (~heritability) among sister species by conducting a correlation analysis for each taxon (e.g., Jablonski 1987; Smith et al. 2004). For each genus containing 2 or more species, a single random species pair was selected and plotted; these represent the analogue of "full sibs" in quantitative genetics. The procedure used minimized the influence of species-rich genera but also discarded much potential information, since not all species were used to derive the regression. Earlier work had demonstrated, however, that results were insensitive to the structuring of the sib-sib regression (Smith et al. 2004). That is, results were similar regardless of whether each species was represented only once (resulting in half as many values as there were species in each genus), all species were paired twice (resulting in as many values as there were species), or all possible pairwise combinations of species within a genus were plotted (which results in a disproportionate influence of specious genera). Because species were chosen randomly and thus were just as likely to be smaller as larger than siblings, variation in the $x$ and $y$ directions was approximately equivalent. The slope of the regression represents an estimate of broad-sense heritability (sensu Falconer 1989). All analyses were repeated for the entire taxon and for a subset of the smallest size classes to examine more carefully patterns among smaller genera.

### *Partitioning of Body Size Variation among the Taxonomic Hierarchy*

We conducted two complementary statistical analyses on the bat, terrestrial mammal, and bird datasets. First, we computed the spatial autocorrelation statistic Moran's *I*, which partitions phylogenetic correlation among taxonomic levels (Gittleman and Kot 1990; Gittleman et al. 1996; Martins and Hansen 1997). The numerator is a measure of covariance among phenotypic traits of species, and the denominator, a measure of variance. Standardized values of Moran's $I$ vary from +1 to −1, with positive values indicating that a trait is more similar than random and negative values indicating they are more different (Gittleman and Kot 1990; Gittleman et al. 1996). Second, we employed mixed-model nested ANOVA to evaluate how much of the variation in body size was attributable to different taxonomic levels. Hence, taxonomic classifications (genus, family, order) were used as the nested independent variables. We used the "Proc Mixed" and "VarComps" procedures in SAS because of the extremely unbalanced design (SAS 1989); mammal analyses were also conducted using an analogous procedure written in an R script.

### *Range and Species Richness of Taxa*

Finally, we derived descriptive statistics for each genus and family. These were tabulated and then plotted separately for each taxon to address several questions about body size diversification: Are larger organisms more variable in body size? Are smaller-bodied genera more likely to contain more species than larger ones, reflecting a greater turnover? How does the ability to fly influence these patterns?

## Results

Our analyses indicate that body size is extremely heritable for all taxa, regardless of whether they are volant (table 8.2). The similarity of body size for congeneric sibling species for birds and terrestrial mammals is ~0.95 for both groups; for bats it is slightly less (~0.86), which may reflect smaller sample size. Further, recall that our sib-sib analysis was conservative, in that we assumed all congeners were equally related and did not incorporate varying branch lengths. Longer evolutionary history would be expected to allow greater diversification and hence result in a reduced estimate of heritability.

Although flight did not influence the pattern of similarity among congeners, there was a difference among taxa. An intriguing pattern found in earlier work was the lack of resemblance in the body size of sister

TABLE 8.2 **Results from Sib-Sib Analysis**

| Taxon | N | Coverage | Equation | $r^2$ | Comment |
|---|---|---|---|---|---|
| Nonvolant mammals | 1,844 | Africa, North America, South America, Australia; restricted to terrestrial mammals | Y = 0.953x + 0.099 | 0.922*** | From Smith et al. 2004 |
| Small nonvolant mammals | 235 | Same as above, but restricted to mammals less than 1.25 log units | Y = 0.223x + 0.706 | 0.113 | From Smith et al. 2004 |
| Volant mammals | 686 | Global | Y = 0.864x + 0.175 | 0.745*** | Present study |
| Small volant mammals | 121 | Global, but restricted to bats weighing less than 0.8 log units | Y = 0.399x + 0.565 | 0.021 | Present study |
| Birds | 5,877 | Global, but restricted to volant birds | Y = 0.959x + 0.071 | 0.919*** | Present study |
| Small birds | 230 | Global, but restricted to birds weighing less than 0.8 log units | Y = 0.953x + 0.086 | 0.431** | Present study |

*Note*: The slope of the regression provides an estimate of broad-sense heritability (Jablonski 1987; Smith et al. 2004). Note the extremely low, nonsignificant values for the smallest body size classes of nonvolant mammals and bats; birds in contrast do not "lose" heritability of body mass at the smallest masses. Sample size and coverage varies somewhat between these studies. Stars indicate significance level: *** = $p < 0.001$; ** = $p < 0.01$.

species for the very smallest mammals (i.e., 0.25 to 1.25 log units; ~1.8 to 18 g; Smith et al. 2004). Heritability estimates decreased from 0.953 to 0.223, and the overall regression became nonsignificant (Smith et al. 2004). When we examine the smallest size classes (those less than 0.8 log units, or ~6.3 g) we find that bats also "lose" heritability at the smallest size classes, with estimates decreasing from 0.864 to 0.399 and the explained variation from 0.745 to 0.021 (table 8.2). Birds, however, display *no* decrease in heritability with smaller body mass (i.e., compare 0.959 for the entire taxon with 0.953 for the smallest size classes). Given that the range of size and the sample size were approximately equivalent for small bats and birds, this result appears to indicate a fundamental difference in the response related to taxonomic affiliation.

The Moran's *I* analysis confirms the conclusions suggested by the sib-sib analysis: all taxa examined demonstrate similar patterns irrespective of their ability to fly. Birds, volant mamals, and nonvolant mammals all show high levels of similarity due to taxonomic relatedness (fig. 8.1). Not only were species within genera significantly *more* similar in body size than would be expected by chance (Moran's $I = 0.822$ for bats, 0.834 for nonvolant mammals, 0.898 for birds; $p << 0.05$ in all cases), but this

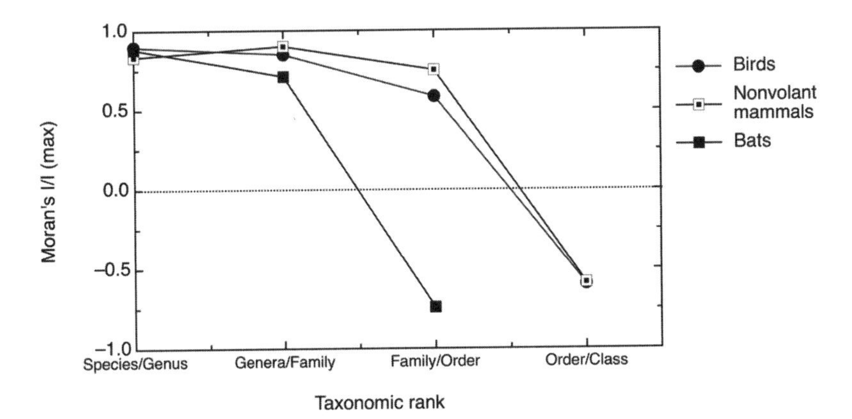


FIGURE 8.1. Phylogenetic autocorrelation at different levels of the taxonomic hierarchy for selected taxa. The standardized Moran's $I$ is plotted for birds (*filled circles*), nonvolant mammals (*open squares*), and bats (*filled squares*). Notice that all groups display the same general trend: the body size of species with genera, and genera within families, is significantly more similar than expected; orders within the class are significantly more divergent than expected. For bats, the dissimilarity is found for families within the order.

pattern was maintained across the taxonomic hierarchy (i.e., genera within families and families within orders). Interestingly, families within the order for bats, and orders within the class for birds and nonvolant mammals, showed the opposite pattern: they were more *dissimilar* in body mass than would be predicted from random (Moran's $I = -0.742$ for bats, −0.729 for nonvolant mammals, −0.596 for birds, $Z = -52.8$, $p << 0.05$; fig. 8.1). These patterns suggest that different bird or mammal orders have distinctive and characteristic body sizes, suggestive perhaps of ecological niche partitioning.

The nested analysis of variance suggested an interesting taxonomic-based difference in the structuring of body size variation. While the majority of variation in mammals is partitioned at the highest levels of the taxonomic hierarchy (e.g., order and family), virtually all of the body size variation is explained at the lowest level for birds (table 8.3; see also fig. 8.2). Generic affiliation is responsible for 93% of the variation in body mass, with the remainder mostly unexplained. Virtually nothing is explained by familial or order affiliations. The constraints of flight do not appear to be involved; bats closely resemble other mammals (table 8.3). Note, however, that because one fewer level was used in the nested ANOVA for bats, more variation is explained at the generic and familial level than for terrestrial, marine, or all mammals together.

TABLE 8.3 **Partitioning of Body Size along the Taxonomic Hierarchy**

| Variance Source | df | Variance Component | Percentage of Total Variance Explained |
|---|---|---|---|
| Birds: | | | |
| Total | 5,876 | 2.65 | 100 |
| Order | 29 | 0.03 | 1.20 |
| Family | 159 | 0.02 | 0.60 |
| Genus | 1,614 | 2.47 | 93.33 |
| Error | 4,074 | 0.13 | 4.86 |
| All mammals: | | | |
| Total | 3,562 | 3.37 | 100 |
| Order | 27 | 2.35 | 69.57 |
| Family | 110 | 0.80 | 23.79 |
| Genus | 936 | 0.18 | 5.32 |
| Error | 2,489 | 0.04 | 1.31 |
| Terrestrial mammals: | | | |
| Total | 2,768 | 2.97 | 100 |
| Order | 25 | 1.78 | 61.97 |
| Family | 83 | 0.85 | 29.59 |
| Genus | 746 | 0.20 | 6.93 |
| Error | 1,914 | 0.04 | 1.51 |
| Bats: | | | |
| Total | 685 | 0.23 | 100 |
| Order | 0 | — | — |
| Family | 14 | 0.09 | 40.39 |
| Genus | 132 | 0.09 | 38.65 |
| Error | 539 | 0.05 | 20.75 |

*Note*: Analysis used a nested-random-effects Analysis of Variance (Mixed Model) from SAS or was written in R script (see text for details).

These results were robust with respect to the type of analysis; when a series of one-way ANOVAs were conducted on genus, family, and order independently, variation was partitioned in a similar fashion.

Interestingly, when nonvolant birds were included in a nested ANOVA, more variation was explained at the ordinal and familial levels than at the genus level (Maurer, chap. 3 above). Since many of these species are found in unique orders and families (e.g., ratites), this result is not surprising. As all modern species of birds are most likely descended from flying species, this implies that when the constraints of flight were removed, significant evolutionary divergence occurred, as lineages exploited ecological opportunities previously unavailable to birds. This is clearly the case for distinctive forms of birds isolated on islands that subsequently lost flight and began to evolve by natural selection to use available resources exploitable by adaptive syndromes characterized by more sedentary modes of locomotion. This is a fundamentally different

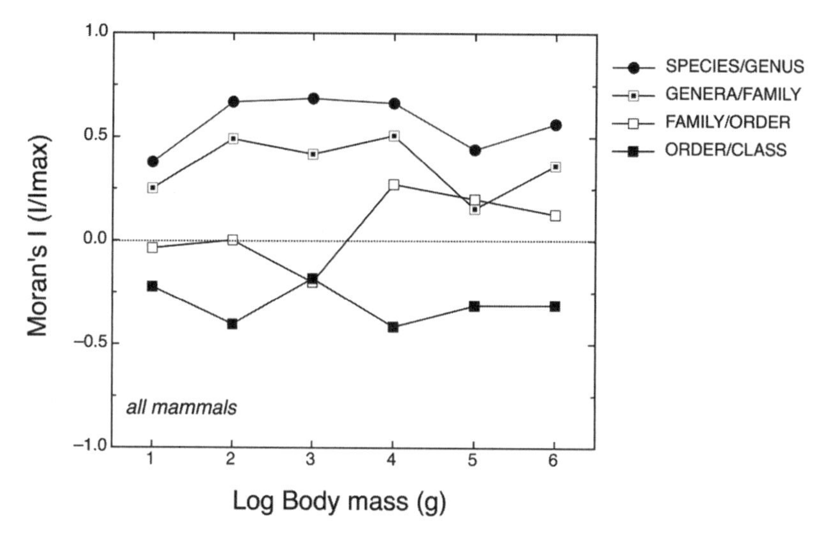


FIGURE 8.2. Phylogenetic autocorrelation for different body sizes of mammals. Shown are the standardized Moran's $I$ values for species within a genus (*filled circles*), genera within families (*open squares with dots*), families within mammalian orders (*open squares*), and orders within the class Mammalia (*filled squares*). Each value represents the mean Moran's $I$ computed for mammals falling within that body size bin; the largest bins are not shown, because sample sizes fell below 100 species for masses greater than 6 log units. Note that in most bins, the variation in body mass is partitioned similarily across the body mass spectrum; this is not the case for the smallest body size bin, which contained mammals ranging from 0.255 to 1.047 log units (N = 381).

sequence of evolutionary events than that which gave rise to bats. Bats represent essentially the opposite sequence: the modification of a primarily nonvolant body plan to allow the exploitation of resources available to flying organisms.

Note that divergent results for birds could arise if ornithologists generally had a more heterogeneous conceptualization of what constitutes a family or order than mammalogists. That is, do scientists specializing on different taxa differ in their use of morphology to classify organisms? This seems unlikely, especially since modern phylogenies incorporate both morphology and genetics. Nonetheless, we tested this idea by repeating our analysis using the Sibley and Alhquist taxonomy (1990). This classification scheme was based on DNA-DNA hybridization studies and has been rejected by a number of ornithologists (especially outside the United States) because it aggregated large sets of birds into single families and proposed a number of unorthodox ordinal arrangements. The Curiae, for example, included birds as diverse as birds of paradise, helmet shrikes, and ioras. Today only portions of this taxonomy are accepted. Interestingly, we found virtually no difference when this taxonomy was employed; again variation was almost completely explained at the generic level (97% in both cases) and similarly partitioned among families and orders in the series of one-way ANOVAs. Thus, we concluded that our results were robust with respect to the taxonomy used, and that birds really do display little or no heritability of body size at successively higher levels of the taxonomic hierarchy. Similarly, although there is a more recent update of Wilson and Reeder, which includes the addition of new species as well as relatively minor reclassifications of already existing ones, we have no reason to expect that it would qualitatively alter our conclusions.

Patterns of evolutionary diversification appeared to be unrelated to the ability to fly and to be unrelated to the body size of taxa. When we examined the number of species within genera and families as a function of size, there was no pattern of increasing species richness with smaller body size for any group (fig. 8.3). This suggests that evolutionary turnover rates are not higher for smaller-bodied species or, alternatively, that higher origination rates are balanced by higher extinction rates. The ability to fly does not appear to result in an increase in the species diversity of genera or taxa, although it does result in upper limitations on the body mass of organisms. We also do not find an increase in the range or variation of body size with genera or families that is related to either the ability to fly or body mass (fig. 8.4). The range is constant across body mass and across taxa. Larger organisms are not more variable in body size, nor are volant organisms more constrained. Most genera and families contain a relatively small amount of body size variation, perhaps because of the high heritability found among taxa (e.g., tables 8.1–8.3; fig. 8.2). Interestingly, despite the limited variation in generic body mass explained by bird families, the range contained is invariant and quite similar to that of mammals (fig. 8.4). Moreover, the range or variation in body size seen within levels of the taxonomic hierarchy does not increase with the species richness of the group (fig. 8.3). This suggests that diversification occurs by "infilling" within the body mass space characteristic of genera, families, and orders.

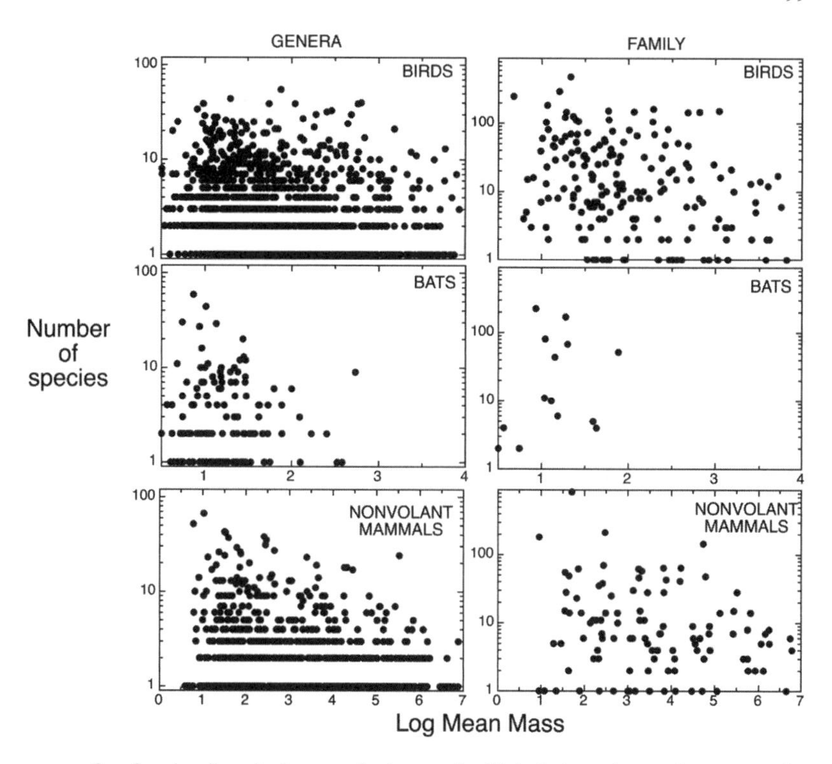


FIGURE 8.3. Species diversity by mean body mass for birds, bats, and nonvolant mammals. Shown is the number of species for each body mass bin for genera (*first column*) and families (*second column*). Patterns are shown separately for each taxon: birds (*first row*), bats (*second row*), and nonvolant mammals (*third row*). Note that species diversity does not increase with body mass for any group at any hierarchical level. Values are log transformed.

## Discussion

Our results clearly indicate that phylogenetic relationships are more constraining than the functional or biomechanical limitations imposed by flight. Despite sharing a major evolutionary innovation with birds, bats demonstrate the same patterns of heritability and body size diversification as other mammals.

Strikingly, the heritability of body size of species within a genus is about 0.95 for all groups regardless of taxonomic or functional affiliation (table 8.2). This has important implications for evolutionary diversification: when species diverge, they generally tend to remain at about the same size as congeners. This may reflect a similar ecological role within communities and/or the scaling of many fundamental physiological, eco-

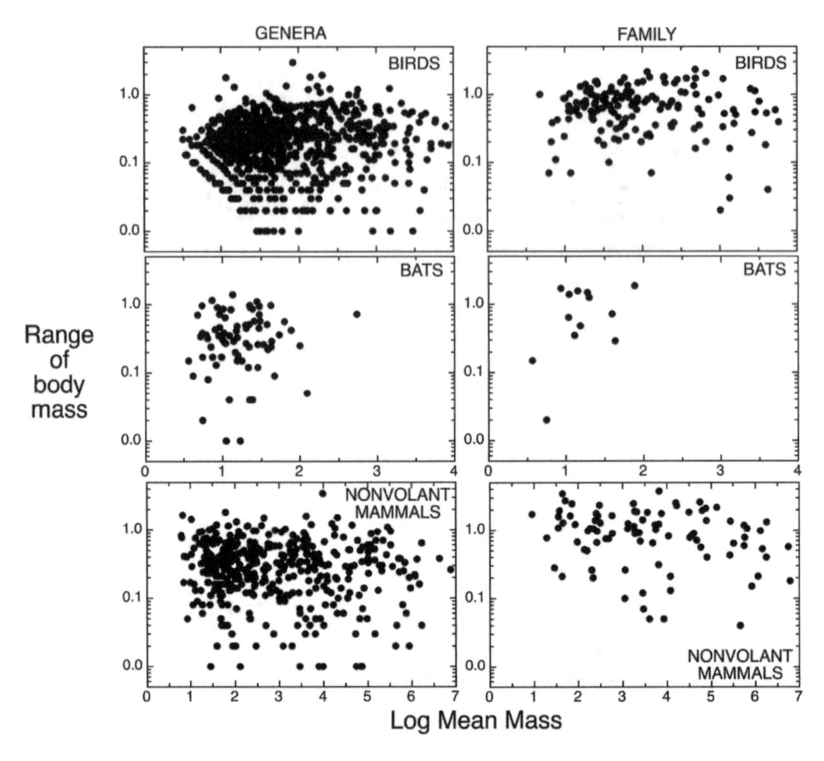


FIGURE 8.4. The range in body mass by mean body mass. Shown is the range in mass for each body mass bin within genera (*first column*) and families (*second column*). Patterns are shown separately for each taxon: birds (*first row*), bats (*second row*), and nonvolant mammals (*third row*). Larger body size does not result in a greater range of body mass within the genus or family. Values are log transformed.

logical, and life history parameters with body mass (Calder 1984). Interestingly, we did not find a pattern between the number of species within a genus or family and body mass for any group (fig. 8.3); smaller-bodied animals did not have more species. Nor was a greater range of mass found in large-bodied genera or families (fig. 8.4). However, while the overall heritability of body mass is the same for both vertebrate groups, we do find two fundamental differences in the patterns between the bats and birds.

First, body size of birds remains highly heritable even for the smallest size classes (table 8.2). This is in sharp contrast to both volant and nonvolant mammals, where the pattern of heritability breaks down completely. At the very smallest body masses, mammalian species within a genus tend to diverge significantly in body mass (table 8.2). Clearly, small body size is subject to a unique set of physiological and life history constraints. Allometric scaling of many fundamental physiological parameters results in increasing mass-specific maintenance costs for animals of decreasing size (i.e., Peters 1983; Calder 1984; Schmidt-Nielsen 1984). We anticipated that flight would exacerbate these constraints and so particularly expected to find similar patterns for small-bodied bats and birds. Indeed, both birds and bats share important physiological adaptations related to flight. For example, bats use oxygen at rates similar to those of birds of the same size rather than other mammals (Kendeigh 1970; Schmidt-Nielsen 1984). Accordingly, bats have a heart size more similar to that of birds than to that of other mammals; its relative size increases with decreasing body mass (Schmidt-Nielsen 1984).

However, our results suggest the mode of diversification may be fundamentally different for the smallest birds and bats: while small mammals appear to evolve divergent body sizes when speciating at smaller sizes, small birds tend to stay at about the same size. Indeed, the heritability of body size is invariant with the size of birds (table 8.2). We have argued elsewhere that for mammals at the extreme end of the body mass spectrum it may be difficult for coexisting species to diverge in traits other than size (Smith et al. 2004). In other words, there is little "physiological space" to develop novel adaptations to new environments, because biomechanical, thermoregulatory, and other constraints limit alterations in body shape that permit ecological specialization in larger species. Given the opportunity to colonize a new and colder ecological niche, small mammals may be unable to increase metabolic rate, insulation or fat reserves, or foraging activities to deal with increased thermoregulatory demands. Instead, they adapt by altering body size. Such energetic constraints might be even more severe because of the cost of flight and its influence on life history characteristics.

One explanation is that birds have more ecophysiological "space" than do bats. While volant birds and bats share a mostly overlapping body size distribution (e.g., Maurer et al. 2004), they appear to diverge in several important ecological and physiological ways. The vast majority of bats are both nocturnal and insectivorous (Fenton 2001). In contrast, volant birds can be nocturnal or diurnal and occupy almost the full spectrum of potential trophic niches, including carnivory, herbivory, and granivory. The greater dimensionality of bird niche breadth implies they may be capable of more fine-scaled niche partitioning than bats, which may explain why after speciation birds remain about the same body size as congeners. If sister species tend to remain in similar environments with similar selective pressures but are able to subdivide the niche space in some finer way than can mammals, there may be no selective pressure for body size to diverge. Additionally, the lower population density of birds compared to mammals (Brown 1995; Maurer 1999) may mean that selective forces operating are weaker.

The reproductive strategies of bats are also markedly different from those of birds. The cost of live birth means female bats carry young with them instead of leaving eggs at the nest. This may reduce mortality of the young, but concomitant biomechanical constraints limit the vast majority of bat species to a single offspring. After parturition, the female is solely responsible for provisioning the offspring through lactation, which is energetically highly demanding (Peters 1983; Schmidt-Nielsen 1984). Moreover, female bats nurse their youngsters until they are nearly adult size, because young bats cannot forage until their wings are fully developed (Fenton 2001). In birds, provisioning of the young is often shared by parents, allowing the production of multiple offspring and clutches in a given year. Thus, to fly, bats have adopted the life history strategies of a much larger mammal, with long life spans and extremely low reproductive output (Calder 1984; Schmidt-Nielsen 1984; Fenton 2001). Birds, however, with more or less the same energetic requirements, have much higher fecundity and generally shorter life spans than bats of comparable body size.

Second, most of the variation in body size in birds is explained at the lowest level of the taxonomic hierarchy (table 8.3). While body size is highly "heritable" for species within genera for all groups, higher-level patterns (e.g., genera within families, and families within orders) only hold true for mammals. Despite this, species within genera, genera within families, and families within orders are significantly more similar in body size than expected by chance for both birds and mammals (fig. 8.2). It is only at the highest taxonomic levels, that of orders within the class, that body size is more dissimilar than expected. This may reflect ecological body mass partitioning. This result is a bit paradoxical, because most of the variation in the body size of birds is explained at the level of species within a genus (~93%; table 8.3), with very little at higher levels of the taxonomic hierarchy. While such a result could indicate taxonomic confusion in birds, we note that use of different taxonomies led to the same conclusion. Thus, despite sharing a major evolutionary innovation and a similar body mass spectrum with birds, bats follow the patterns found for other mammals in all respects. Although functional constraints do influence the overall similarities in the body mass distributions of birds and bats (Maurer et al. 2004), we do not find convergence in other aspects of body size diversification.

Interestingly, our results are consistent with recent findings by McGowan and Dyke (2007) indicating a clear separation in the "ecomorphospace" of birds and bats. These authors suggested that having the wing attached to the hind limb constrained the morphological diversity of bats relative to that of birds. The lack of overlap in ecomorphospace was maintained after the extinction of the pterosaurs, suggesting that it was not the result of competitive exclusion. This is not to suggest that bats are competitively inferior to birds; rather, bats are highly derived specialists that have evolved an alternative pathway to flight (Swartz and Middleton 2008). The wings of bats are much thinner than those of birds, so bats are capable of quick and agile maneuvers. However, the mechanical properties of the bones of the wings and the loads experienced by the bat skeleton during flight are different from those of other volant groups (Swartz and Middleton 2008).

Our results suggest that although the ability to fly has required body plan modification for bats and a series of common ecological constraints shared with birds, it has not led to common evolutionary constraints (Niklas 2000; Price 2003). Instead, what appears to constrain evolution at the lower end of the body size spectrum for bats is that they are small mammals and not that they fly. Bats are not furry birds, but rather flying rats.

## References

Alexander, R. M. N. 2003. *Principles of animal locomotion.* Princeton: Princeton University Press.

Brown, J. H. 1995. *Macroecology.* Chicago: University of Chicago Press.

Calder, W. A. 1984. *Size, function, and life history.* Cambridge, MA: Harvard University Press.

Campbell, K. E., Jr., and E. P. Tonni. 1980. "A new genus of teratom from the Huayquerian of Argentina (Aves: Teratornithidae)." *Contributions in Science, Natural History Museum of Los Angeles County* 330:59–68.

Clements, J. F. 1991. *Birds of the world: A checklist.* 4th ed. Vista, CA: Ibis Publishing.

——. 2007. *The Clements checklist of birds of the world*. 6th ed. Ithaca: Cornell University Press.

Dunning, J. B. 1993. *CRC handbook of avian boody masses*. Boca Raton, FL: CRC Press.

Falconer, D. S. 1989. *Introduction to quantitative genetics*. London: Longman.

Fenton, B. M. 2001. *Bats*. New York: Checkmark Books.

Gittleman, J. L., C. G. Anderson, M. Kot, and H. K. Luh. 1996. "Comparative tests of evolutionary lability and rates using molecular phylogenies." In *New uses for new phylogenies*, edited by P. H. Harvey, A. J. Leigh Brown, J. Maynard Smith, and S. Nee, 289–307. Oxford: Oxford University Press.

Gittleman, J. L., and M. Kot. 1990. "Adaptation: Statistics and a null model for estimating phylogenetic effects." *Systematic Zoology* 39:227–241.

Heaney, L. R., and P. D. Heideman. 1987. "Philippine fruit bats: Endangered and extinct." *Bats* 5 (1): 3–5.

Jablonski, D. 1987. "Heritability at the species level: Analysis of geographic ranges of Cretaceous mollusks." *Science* 238 (4825): 360–363.

Jones, K. E., and A. MacLarnon. 2001. "Bat life-histories: Testing models of mammalian life history evolution." *Evolutionary Ecology Research* 3: 465–476.

Kendeigh, S. C. 1970. "Energy requirements for existence in relation to size of bird." *Condor* 72 (1): 60–65.

Marden, J. H. 1994. "From damselflies to pterosaurs: How burst and sustainable flight performance scale with size." *American Journal of Physiology* 266 (4): R1077–R1084.

Martins, E. P., and T. F. Hansen. 1997. "Phylogenies and the comparative method: A general approach to incorporating phylogenetic information into the analysis of interspecific data." *American Naturalist* 149 (4): 646–667.

Maurer, B. A. 1999. *Untangling ecological complexity: The macroscopic perspective*. Chicago: University of Chicago Press.

Maurer, B. A., J. H. Brown, T. Dayan, B. J. Enquist, S. K. M. Ernest, E. A. Hadly, J. P. Haskell, D. Jablonski, K. E. Jones, D. M. Kaufman, S. K. Lyons, K. J. Niklas, W. P. Porter, K. Roy, F. A. Smith, B. Tiffney, and M. R. Willig. 2004. "Similarities in body size distributions of small-bodied flying vertebrates." *Evolutionary Ecology Research* 6 (6): 783–797.

McGowan, A. J., and G. J. Dyke. 2007. "A morphospace-based test for competitive exclusion among flying vertebrates: Did birds, bats and pterosaurs get in each other's space?" *Journal of Evolutionary Biology* 20 (3): 1230–1236. doi: 10.1111/j.1420-9101.2006.01285.x.

Meunier, K. 1951. "Korrelation und Umkonstruktionen in den Größenbeziehungen zwischen Vogelflügel und Vogelkörper." *Biologia Generalis* 19:403–443.

Niklas, K. J. 2000. "The evolution of plant body plans: Biomechanical perspective." *Annals of Botany* 85 (4): 411–438.

Pennycuick, C. J. 2008. *Modelling the flying bird*. New York: Academic Press.

Peters, R. H. 1983. *The ecological implications of body size*. Cambridge: Cambridge University Press.

Price, P. W. 2003. *Macroevolutionary theory on macroecological patterns*. Cambridge: Cambridge University Press.

SAS. 1989. *SAS/STAT user's guide, version 6*. 4th ed. Cary, NC: SAS Institute.

Sato, K., K. Q. Sakamoto, Y. Watanuki, A. Takahashi, N. Katsumata, C. Bost, and H. Weimerskirch. 2009. "Scaling of soaring seabirds and its implication for the maximum size of flying pterosaurs." *PLoS ONE* 4 (4): e5400. doi: 10.1371/journal.pone.0005400.

Schmidt-Nielsen, K. 1984. *Scaling, why is animal size so important?* Cambridge: Cambridge University Press.

Sibley, C. G., and J. E. Ahlquist. 1990. *Phylogeny and classification of birds*. New Haven: Yale University Press.

Smith, F. A., J. H. Brown, J. P. Haskell, S. K. Lyons, J. Alroy, E. L. Charnov, T. Dayan, B. J. Enquist, S. K. M. Ernest, E. A. Hadly, K. E. Jones, D. M. Kaufman, P. A. Marquet, B. A. Maurer, K. J. Niklas, W. P. Porter, B. Tiffney, and M. R. Willig. 2004. "Similarity of mammalian body size across the taxonomic hierarchy and across space and time." *American Naturalist* 163 (5): 672–691.

Smith, F. A., S. K. Lyons, S. K. M. Ernest, K. E. Jones, D. M. Kaufman, T. Dayan, P. A. Marquet, J. H. Brown, and J. P. Haskell. 2003. "Body mass of late Quaternary mammals." *Ecology* 84 (12): 3403–3403.

Swartz, S. M., and K. M. Middleton. 2008. "Biomechanics of the bat limb skeleton: Scaling, material properties and mechanics." *Cells Tissues Organs* 187 (1): 59–84. doi: 10.1159/000109964.

Templin, R. J. 2000. "The spectrum of animal flight: Insects to pterosaurs." *Progress in Aerospace Sciences* 36 (5–6): 393–436.

Vizcaíno, S. F., and R. A. Fariña. 1999. "On the flight capabilities and distribution of the giant Miocene bird *Argentavis magnificens* (Teratornithidae)." *Lethaia* 32 (4): 271–278.

Wilson, D. E., and D. M. Reeder. 1993. *Mammal species of the world: A taxonomic and geographic reference*. 2nd ed. Washington, DC: Smithsonian Institution Press.

Zann, R. A. 1996. *The zebra finch: A synthesis of field and laboratory studies*. Oxford: Oxford University Press.

CHAPTER NINE

# On Body Size and Life History of Mammals

James H. Brown, Astrid Kodric-Brown, and Richard M. Sibly

"Live fast, die young, and leave a good looking corpse" —*Knock on Any Door* (movie, 1949)

"The positive allometry of horns and antlers is one of the most pervasive and least understood regularities in the study of form and function" —S. J. Gould (1974)

## Historical Introduction

Mammals vary enormously in body size. An elephant is about 100,000 times larger than a mouse, but the size range across all mammals is much greater, a factor of $10^7$, or 10,000,000 times. The smallest mammals are shrews weighing about 2 g, whereas the largest is the blue whale weighing 200 tons. A shrew is the size of a marble; a whale is the size of a house. Mammals have been around for about 200 million years. During that time their ancestors diverged from reptiles, survived the asteroid impact that caused the extinction of the dinosaurs 65 million years ago, and then radiated and diversified. Today they number about 5,000 species, representing some 1,200 genera, 153 families, and 29 orders.

Contemporary mammals differ enormously in form, function, and ecology. They include shrews and mice, which are tiny and very active, foxes and cats, which hunt the shrews and mice, buffaloes and zebras, which graze in grasslands, monkeys and sloths, which are adapted to life in trees, and elephants and whales, which are huge and slow. Despite the obvious differences in body size, shape, and lifestyle, however, all mammals except for the egg-laying monotremes share many attributes of anatomy and physiology, including hair, endothermy, mammary glands, and live birth. For at least the last two centuries, studies of mammals have emphasized these two themes: fundamental similarities reflecting common ancestry and body plans and substantial differences reflecting evolutionary divergence and ecological adaptation.

By the 1930s investigators were beginning to show how the study of body size scaling relations could contribute to understanding the shared and divergent traits. Thompson (1917), Huxley (1932), and others demonstrated that many attributes of animals varied with body size, and that often the variation was so precise that it could be described by a mathematical equation, a power law of the form

$$Y = aX^b, \tag{1}$$

where $Y$ is some dependent variable, such as heart rate, litter size, or life span, $X$ is some measure of body size, usually mass, $a$ is a normalization constant that typically varies with phylogenetic or functional group and sometimes with environmental setting, and $b$ is another constant, the allometric or scaling exponent. Huxley (1932) coined the term "allometry," meaning different scale, for biological scaling relationships described by equation 1. Huxley (1932) and Thompson (1917) showed that during ontogenetic growth organisms do not change shape isometrically or geometrically like simple physical objects such as spheres, cubes, or nested Russian dolls. As mammals grow from newborn to adult, some body parts, such as the head and brain, decrease in relative size, whereas others, such as the antlers and horns of deer and antelopes, increase differentially.

In the early 1930s, independent studies by Kleiber (1932) and Brody and Proctor (1932) showed that metabolic rates of adult mammals of different species scale with body mass with an exponent very close to ¾. This observation, which has subsequently been termed Kleiber's law, was surprising, because metabolic rate might be expected a priori to scale with body mass with an exponent of either 1, because energy required to support a body of living tissue should scale linearly with mass, or ⅔, because mammals, being endothermic, must dissipate the heat generated by metabolism from their body surface, which scales with the

⅔ power of mass. Over the next few decades, many studies of biological rates and times revealed that they also scaled with distinctive exponents that typically were very close to simple multiples of ¼. So, for example, most biological rates, such as heart, respiratory, and reproductive rate of mammals, scale as $M^{-1/4}$, whereas biological times, such as cell cycle time, gestation time, and life span, scale as $M^{1/4}$.

In the early 1980s four excellent books (McMahon and Bonner 1983; Peters 1983; Calder 1984; Schmidt-Nielsen 1984) reviewed and synthesized these studies of size-related variation. Although each of these books examined the correlates and consequences of body size from a somewhat different perspective, together they eloquently summarized the state of the science at the time. They unanimously concluded that most biological scaling relations were characterized by quarter-power exponents. They advanced this conclusion as an empirical generalization, however, because they were unable to provide a general mechanistic explanation for it.

In the absence of a unifying theory, research on animal allometry stagnated. Then, in the late 1990s, West et al. (1997, 1999) published a new theory, which showed how $M^{3/4}$ scaling of whole-organism metabolic rate and many other quarter-power scaling relations emerged naturally from biophysical models of vascular systems. They suggested that the origin of the quarter powers lay in the "additional linear dimension" introduced by the hierarchical fractal-like designs of the networks within the bodies of organisms which distribute the resources that fuel cellular metabolism. This work remains controversial, but it is stimulating renewed interest in biological scaling relations. It is leading to a unified conceptual framework for comparative biology that we call metabolic scaling theory.

One area where the theory is being applied is life history. The energy and materials of biological metabolism are allocated to maintenance, growth, and reproduction. Since Darwinian fitness depends on these three components of the life history, the scaling of metabolic rate powerfully constrains the allocation of resources to these essential activities. In particular, in order to be fit and leave descendants, a mammal must not only maintain its existing biomass, including defending itself from predators, parasites, and diseases. It must also produce new biomass and allocate it between growth and reproduction. In what follows we will apply metabolic scaling theory to explore how several aspects of mammal life history—namely, production rates of females and body sizes, sexual ornaments, and weapons of males—vary with body size.

## Scaling of Female Production

Mammals are ideal for investigating the scaling of biomass production, its allocation to growth and reproduction in males and females, and its influence on life history evolution. The metabolism of the female parent supplies virtually all of the energy and materials that go directly into production of offspring. The male parent typically supplies only a sperm, which contains half of the genetic material but only an infinitesimal fraction of the energy and nutrients in the fertilized egg. Moreover, with internal fertilization, gestation in the uterus, and nursing of offspring during lactation, the mother provides all of the nutrition to rear offspring from conception to weaning, which is about ⅓ of adult body mass in most taxa and nearly equal to adult mass in bats. Additionally, in many species the mother provides much of the energy for thermoregulation with her own body heat and by constructing an insulating nest or burrow. The mother also expends energy to reduce predation on her young, by carrying them in her body prior to birth and by a number of behavioral adaptations for protecting dependent young after birth. In some species, males provide parental care, and in some social mammals, such as mole rats, meerkats, African wild dogs, and wolves, members of the colony provide important contributions.

### *Theory*

If metabolism fuels production, and mass-specific metabolic rate scales with mass as $M^{-1/4}$, it follows that mass-specific production rate of female mammals should scale as $M^{-1/4}$ (Peters 1983; Calder 1984; Charnov 1993; Brown et al. 2004). Ecologists traditionally have measured the production/biomass ratio for a population, where production is the rate of production of new biomass, and biomass is the average standing biomass of the population, both usually measured on an annual basis. This, too, scales as $M^{-1/4}$ in mammals (Peters 1983). Given the unique features of mammalian life history—where the mother supplies energy during gestation and lactation, rears offspring to relatively large size, and protects them from predation and abiotic conditions—the rate of mass-specific production is closely related to the birth rate, which also scales as approximately $M^{-1/4}$ (Brown and Sibly 2006) since weaning mass is close to a fixed fraction of adult mass.

It necessarily follows that death rate must scale similarly, and hence also as approximately $M^{-1/4}$. To see why this must be so, consider two lines of reasoning. First, most species of mammals are neither increasing exponentially nor decreasing rapidly toward extinction. Their populations are remaining relatively stable, or at least fluctuating around some number. This means that long-term population growth rate averages close to zero, and birth rate approximately equals death rate. The second argument is an evolutionary one. Suppose a few individuals within a population acquire a heritable trait that increases fitness by increasing the birth rate. These individuals will tend to leave more offspring, and the population will tend to grow. Eventually, however, density-dependent ecological compensation due to resource limitation or some other process will kick in, and further population increase will be prevented by a compensatory increase in death rate.

So a prediction from metabolic scaling theory is that rates of mass-specific production, birth, and death all scale negatively with body mass, as approximately $M^{-1/4}$. This is shown diagrammatically in figure 9.1. It is a formal quantification of the popular adage that "small animals live fast and die young." It has four important consequences. First, it means that the allometries impose a necessary trade-off between survival and reproduction. Since birth and death rates both decrease with body size as $M^{-1/4}$, average life span or survival time must scale as the reciprocal of these rates and increase with body mass as $M^{1/4}$. Second, the total mass-specific production over a lifetime must be approximately invariant. Lifetime production equals mass-specific production rate per unit time times the life span, so $M^{-1/4}M^{1/4} = M^0$ (Charnov 1993; Charnov et al. 2007).

Third, the approximately $M^{-1/4}$ scaling of production rate means that if everything else including death rate were equal, small mammals would have an enormous advantage, because they produce descendant offspring at a much higher rate. Under ideal conditions a 20 g female mouse can theoretically produce about 20 times her own body mass in offspring in one year, whereas a 2 ton elephant can only produce a fraction of her mass in a year. Such rates are often expressed in doubling times, which would give about 60 days for the mouse and years for the elephant. But of course, as shown above, death rates are not equal across all mammals; they scale with body size the same as birth rates. Whatever theoretical advantages accrue to small mammals by virtue of their high rates of birth and production must be matched in large mammals by counterbalancing advantages of low death rates and long life spans. Out in the real

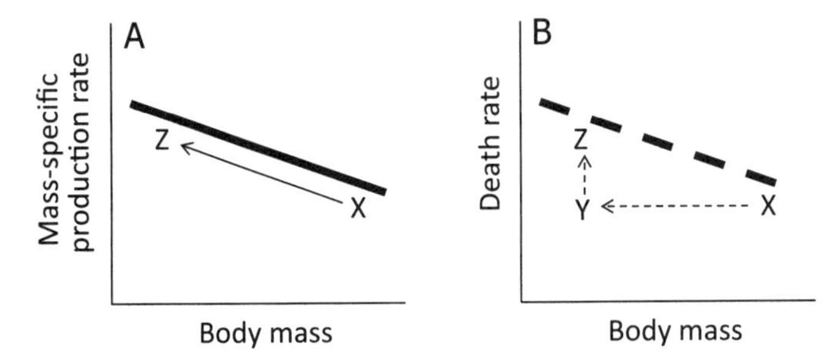


FIGURE 9.1. Diagrammatic representation of life history predictions from metabolic scaling theory in log-log plots. Mass-specific production and death rates are predicted to scale negatively with body mass, as approximately $M^{-1/4}$, as depicted by the heavy solid and dashed lines. Suppose a mutation or a change in lifestyle produces an evolutionary change to smaller bodies with higher rates of production, as shown by the solid arrow $X \rightarrow Z$ in *A*. Initially there may be no change in death rate, as shown by the dashed arrow $X \rightarrow Y$ in *B*. With increased production rate but unchanged death rate the species initially increases, but ecological compensation will eventually kick in and return the population close to the allometric scaling line (Y → Z in *B*). After (Brown and Sibly 2006).

world large and small mammals coexist and persist. Neither size has an advantage or it would have taken over the world.

Fourth, the above theoretical considerations provide a foundation for understanding life history evolution in mammals. It is easy to see how species subject to the same metabolic constraint can move back and forth along the production and death rate allometry lines in figure 9.1. A decrease in body size will tend to result in increased rates of production and birth. Such a change will be favored by natural selection unless there is a matching increase in death rate. The advantage will be only temporary, however, because density-dependent ecological processes will eventually kick in, and death rate will rise to match the birth rate and return the population to the constraint line. Similarly, an increase in body size will tend to result in decreased death rate. This may confer a temporary advantage, but eventually compensation will occur and birth rate will decrease and come to equal death rate. So it is easy to see why, as new lineages have originated, speciated, and diversified, the descendant species have become distributed along the allometric constraint lines.

Figure 9.2 plots data on production rates of real mammals. The predicted negative allometry is obvious, but it is also obvious that not all species lie close to the overall production constraint line. There is con-

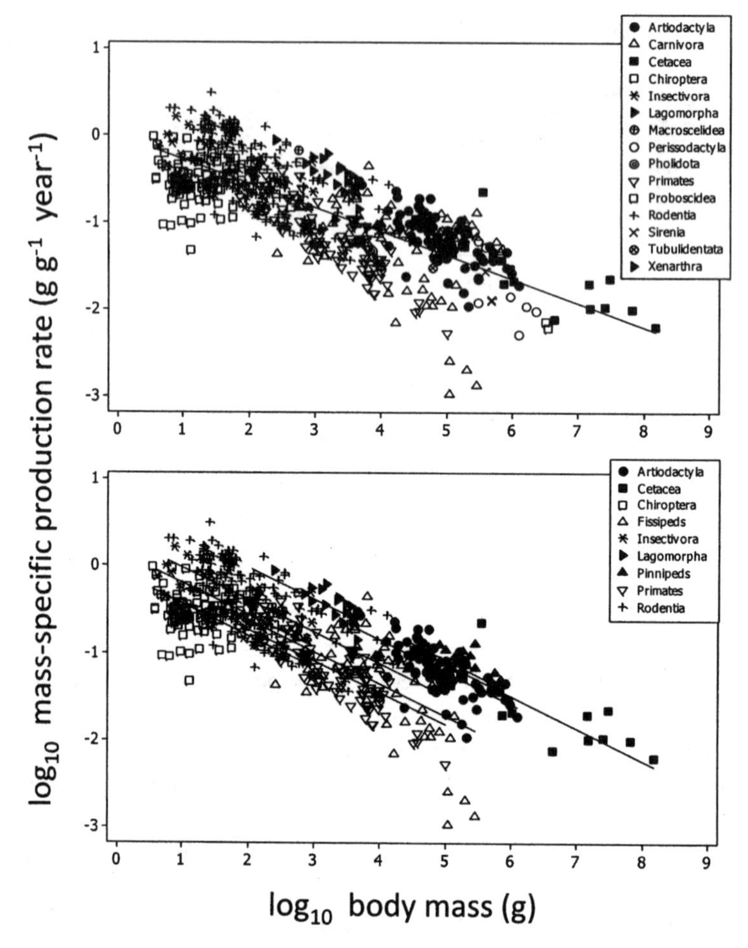


FIGURE 9.2. Production rates of 637 species of placental mammals. Production rates are here estimated as annual production of reproductive biomass divided by female mass (see text for further details). The line in the upper graph was fitted by regression to all the data. The parallel lines in the lower graph were fitted by general linear model (GLM) to selected taxonomic groups as shown. After Sibly and Brown (2007).

siderable variation. Mammals of the same body size differ by about tenfold in their rates of production. Such variation implies that certain kinds of evolutionary changes allowed populations to shift production rates so as to remain permanently above or below the original constraint line. How can this occur? Imagine an evolutionary innovation that allowed some members of a population to achieve a temporary increase in production rate. Below, we will discuss the kinds of traits that might have such an effect. For the moment, however, simply note that the high rate of production could theoretically be sustained so long as the population could persist in the face of the increasing death rates that would occur once density-dependent compensatory processes kicked in. Similarly, an evolutionary change that reduced death rate would persist so long as the population could sustain the concomitant compensation-induced decrease in birth and production rates.

The metabolic framework developed here suggests a somewhat different perspective on life history theory. Traditionally, life history theory has focused on trade-offs in the details of fecundity, mortality, and parental care, without considering explicitly their fundamental basis in energetics and scaling. Metabolic theory suggests that the near-universal $M^{3/4}$ scaling of metabolic rate places a fundamental allometric constraint on mass-specific production rate, causing it to scale as $M^{-1/4}$, and the scaling of life history rates and times follow, scaling as $M^{-1/4}$ and $M^{1/4}$, respectively. So body size is the most fundamental axis of life history variation. Residual variation around or orthogonal to the metabolic constraint reflects the influence of other factors that allow mammals of similar body size to have substantially different life histories. Having provided a theoretical context, we will now use the data on mammals to evaluate hypotheses about the roles of these factors in the ecology and evolution of life history traits.

### *Data and Analysis*

To compare production across species of different body size and lifestyles, it is convenient to use the rate of production per unit body mass per year. This rate, $P$, can be estimated for mammals as the product of the mass of an average offspring ($M_O$) times the average number of offspring per litter ($N_L$), times the average litters per year ($L_Y$), all divided by the mass of an average female ($M_F$), so

$$P = \frac{M_O N_L L_Y}{M_F}. \quad (2)$$

Estimating the rate of production in this way allows us to compile data for hundreds of mammal species for which the relevant life history data have been obtained. We use three datasets: (1) for 532 species of nonvolant placental mammals from Ernest (2003); (2) for 105 species of bats from K. Jones (pers. comm.); and (3) for 237 species of marsupials and 3 species of monotremes from Hamilton et al. (2011). Use of these datasets to calculate production rates requires us to make several simplifying assumptions. First, our calculation does not include all lifetime production. Most notably missing is growth of offspring after weaning, which is fueled by their own metabolism. Second, in making these estimates, we have used the average mass of an offspring at birth, only partway through the period of maternal investment. It would be preferable to use the mass of offspring and size of litter at the end of maternal investment and hence at weaning, but these data were available for only a fraction of the species in our dataset. Third, our data are expressed on an annual basis, which may introduce some errors, especially in small mammals that may live for less than one year and have a reproductive period of only a few months. Finally, our dataset allows estimation of production of a reproductive female, so it does not include any contributions of males or members of social groups in caring and providing for the young.

In part because of these simplifying assumptions, we regard our results and conclusions as preliminary. We are confident that the patterns shown below are qualitatively robust, but the quantitative details may change when more data become available and more rigorous analyses can be performed. Future analyses may also benefit from using other statistical methods. In particular, the data that we present are for species. To obtain preliminary insights into the roles of phylogenetic constraints and adaptive innovations, we have combined species into taxonomic or functional groups. As more complete phylogenetic reconstructions become available, it will be desirable to conduct analyses designed to focus explicitly on the influences of phylogeny and ecology. For the moment, however, we use the term "lifestyle" to refer to a suite of adaptations for a particular way of living. By grouping species according to lifestyle, we can begin to see how both phylogenetic constraints and ecological adaptations have shaped mammalian life histories.

### *Inference: Lifestyles of Eutherian (Placental) Mammals*

A plot of mass-specific production rate as a function of body mass on logarithmic axes for all 637 eutherian mammal species is shown in figure 9.2. Several things are immediately apparent. First, there is a strong negative relationship between production rate and body size. The value of the exponent is −0.28, slightly lower (more negative) than the theoretically predicted value of −¼. (Because production rate is defined as per unit body mass, plotting this measure as a function of body mass has the disadvantage that ratios of random numbers regressed against their denominator automatically yield negative correlations. We nonetheless use this ratio for several reasons: [1] The problem of spurious correlation is not serious when the correlation between numerator and denominator is high, as it is in our data. [2] The slope was less than that for whole-organism production rate by exactly 1, as expected. [3] Using mass-specific values did not affect our calculated estimates of coefficients, standard errors, and significance levels of other variables. [4] Using mass-specific values allows ready biological interpretation and direct comparisons with previous work, without compromising testing of hypotheses [see Meiri et al. 2011].) Second, body size accounts for the majority, 59%, of the variation, but by no means all of it. Mammals of the same size can differ in production rate by up to two orders of magnitude. Third, the variation is related to taxonomy and lifestyle. Certain orders and some taxonomic and functional groups within orders have consistently high or low production rates relative to other mammals.

To analyze the production data more quantitatively we have fitted the data using a general linear model (GLM) procedure. Results are shown in the lower graph of figure 9.2. The GLM procedure fits parallel lines with a common slope to the different groups. The value of this exponent is −0.37, significantly more negative than the predicted −¼. We will not discuss this discrepancy further except to say that we find a similar pattern in marsupials, and we tentatively attribute the consistent deviations from quarter-power scaling to fairly modest but consistent scaling of preweaning mortality (see Hamilton et al. 2011).

By fitting parallel lines to the different groups, the GLM calls attention to the normalization constants or intercepts. These differ by about one order of magnitude, with bats and primates having the lowest values, insectivores and carnivores intermediate ones, and lagomorphs, artiodactyls, pinnipeds, and cetaceans the highest values. So the patterns suggest that production rates of mammals vary along a second axis that is orthogonal to body size and is related to taxonomy and hence to phylogeny. Indeed, the intermediate values of the insectivores and carnivores probably approximate the ancestral condition for eutherian mammals, and the lower or higher rates of other lineages represent evolutionary divergence. The patterns also suggest that the second axis is related to lifestyle. Flying bats and arboreal primates have low rates, whereas marine mammals and folivorous lagomorphs and artiodactyls have high ones (lower graph of fig. 9.2).

This variation can be explained in terms of adaptive changes in birth and death rates, as developed above. We hypothesize that the low production rates of bats and primates reflect life history trade-offs associated with reduced mortality due to predation. When ancestral primates became arboreal and ancestral bats evolved flight, they evaded many predators of terrestrial mammals. So these groups benefited from reduced death rates. But birth rates and death rates have to balance, so through a process of density-dependent ecological compensation, birth rates decreased until they matched the reduced death rates. The reduced production rates of these two groups are not disadvantageous, because they are compensated by reduced death rates, made possible by their unique lifestyles. We hypothesize that the high production rates of folivorous and marine mammals reflect their exploitation of abundant food resources. Morphological, physiological, and behavioral adaptations of these groups allow access to abundant food, which can be readily harvested to fuel high rates of production. Increased rates of offspring production gave the ancestors of these groups an initial advantage, but ecological compensation offset such gains by increased death rates.

If these hypotheses have merit, they lead naturally to corollary predictions that can be evaluated with the dataset. First, within mammalian orders, subgroups and even individual species that have evolved divergent lifestyles should also exhibit divergent life histories. So, for example, lifestyles that confer lowered mortality due to predation should also lead to lower production rates than the lifestyles of closely related mammals. Among the rodents, arboreal squirrels and the several independently evolved groups of fossorial and desert rodents would be predicted to have lower production rates than other rodents due to lower risk of predation, whereas folivorous rodents would be predicted to have higher rates due to their abundant food supply. The data in figure 9.3*B* support this prediction. Similarly, mammals that have escaped most of their predators by evolving very large size would be expected to have low production rates. Supporting this prediction, the largest bears have much lower production than other carnivores (fig 9.3*A*), and the megaherbivore elephants (Proboscidea) and rhinos (largest Perissodactyla) have lower rates than smaller folivorous mammals (upper graph in fig. 9.2). Lifestyles that provide access to abundant food should often allow high

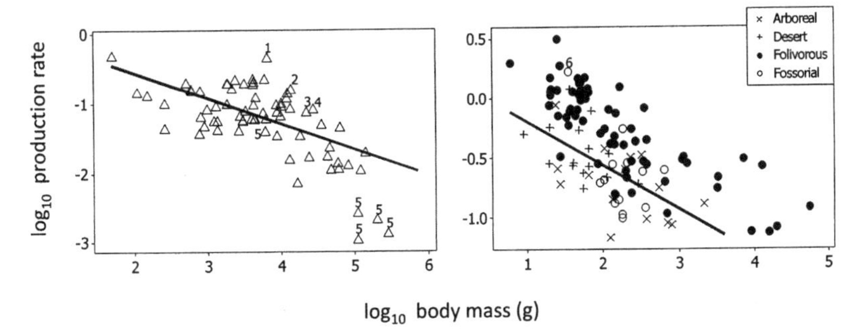


FIGURE 9.3. Production rates of selected groups. *A*, Terrestrial carnivores; *B*, rodents. In each panel, for purposes of comparison, we show the line for insectivores in the lower panel of figure 9.2. Some exceptional cases are highlighted: 1, crab-eating fox (*Cerdocyon thous*); 2, African civet (*Civettictis civetta*); 3, sea otter (*Enhydra lutris*); 4, African wild dog (*Lycaon pictus*); 5, bears of the family Ursidae; 6, naked mole rat (*Heterocephalus glaber*). See text for discussion. After Sibly and Brown (2007).

rates of production. So it is probably not coincidental that two unusual species that have specialized on marine food resources, the sea otter and crab-eating fox, have higher production rates than most fissiped carnivores. Two other species with high production rates are the African wild dog and naked mole rat, highly social species whose colony members supply some of the production to help the reproductive female to rear her offspring (fig. 9.3).

One interesting feature of mammal life history evolution is the relationship between phylogeny and lifestyle. On the one hand, all of the species in several mammalian orders (Chiroptera, Primates, Artiodactyla, Pinnipedia, Cetacea) appear to have substantially lower or higher production rates than most other mammals. This suggests that the divergent life histories that accompanied the ancient divergence of these lineages have been preserved in their descendants, because they continued to have similar conservative niches or lifestyles. On the other hand, multiple lineages that have evolved convergent lifestyles, such as fossorial rodents, have concomitantly evolved convergent life history traits. Individual species, such as the sea otter and crab-eating fox, which have diverged recently to exploit new abundant food resources, have also diverged simultaneously in their production rates. Both phylogeny and ecology have influenced the evolution of mammalian life histories.

So far we have been concerned with the overall rate of production of reproductive biomass. But resources channeled into reproduction are

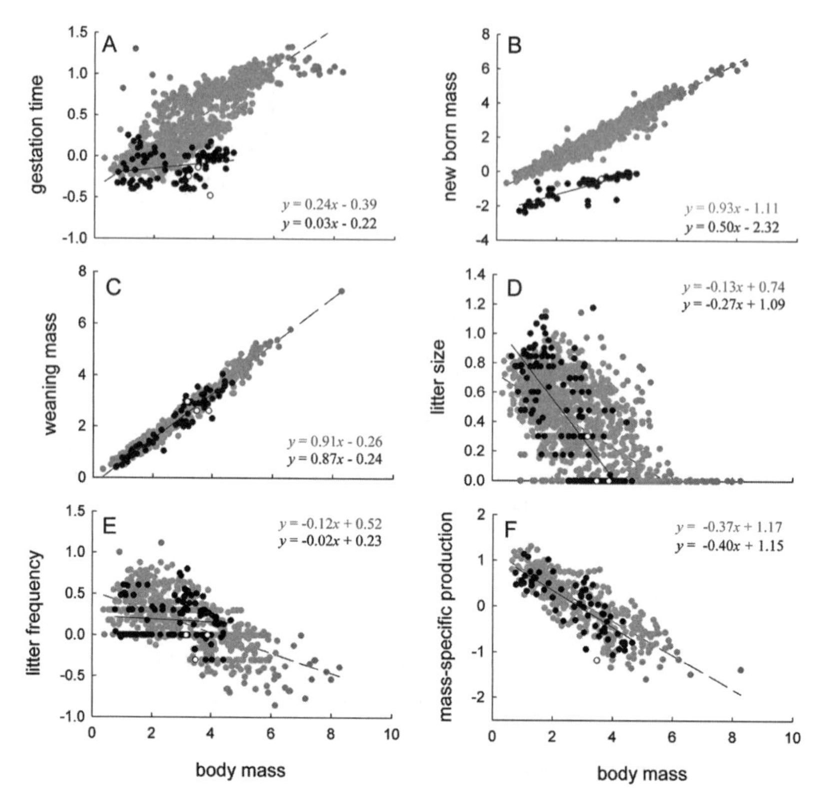


FIGURE 9.4. Comparisons of selected life history traits of placentals (*gray circles*), marsupials (*black circles*), and monotremes (*open circles*) using logarithms to the base 10 on all axes. Life history timings are in months; body masses in grams. Note that although some components of the life history differ conspicuously between placentals and marsupials, data for overall mass-specific production overlap completely and exhibit a common scaling relationship that presumably reflects fundamental metabolic constraints.

also divided up among offspring and litters, and there is much variation in such allocation. Figure 9.4*D* and *E* shows the overall variation between species in litter size and frequency. The main factor determining litter size and frequency is the rate of production of the mother, which as we have seen depends on body size and lifestyle. Small species have higher rates than large ones, because they have some combination of more offspring per litter and more frequent litters. The most productive small mammals, which include some rodents, insectivores, and lagomorphs, can have ten or more offspring per litter and six or more litters per year. By contrast, many large species, including most megaherbivores and marine mammals, give birth to only a single offspring at intervals of a year or more.

As indicated above, mortality strongly influences production rate, albeit indirectly, because in the long term rates of birth and death must balance. So we noted that mammals that lead generally riskier lives, such as terrestrial compared to arboreal and fossorial rodent species, have higher rates of production. Here we note that predation on offspring appears to be a major factor determining how female placental mammals allocate production between a few large and many small offspring within a litter and between a few large litters and many small ones within a reproductive season (Sibly and Brown 2009). Insectivores, fissiped carnivores, rabbits (among the lagomorphs), and most rodents, whose offspring are protected in burrows or nests, produce large litters of small newborns. Artiodactyls, perissodactyls, cetaceans, and hares (among the lagomorphs), which give birth in the open on land or in the sea, produce a few large precocial offspring, because this increases their chances of escaping predation. Similarly, primates, sloths, and anteaters, which carry their young from birth until weaning, also produce litters of one or a few offspring.

It has long been recognized that the substantial variation in mammalian life histories is related to body size and other factors. In particular, the life history traits have been portrayed as being distributed along a fast-slow continuum, so that some mammals, especially small ones, such as folivorous voles, live fast and die young, whereas others, such as large primates, megaherbivores, and whales, live slowly and to advanced ages (e.g., Harvey and Bennett 1983; Harvey and Clutton-Brock 1985; Read and Harvey 1989; Promislow and Harvey 1990; Harvey and Nee 1991; Harvey et al. 1991; Purvis and Harvey 1997; Ross 1998). Other authors have emphasized the influence of body size on the allometric scaling of life history traits (Blueweiss et al. 1978; Peters 1983; Calder 1984; Charnov 1991, 1993). McNab (1986, 2002) has pointed out the relation between lifestyle and metabolic rate, which must fuel production. So much of what is presented above is not new, just old ideas and data repackaged in a somewhat different way. We hope, however, that this theoretical and empirical exploration of the scaling of mammalian production rates, by calling attention to two main orthogonal axes of life history variation, body size and lifestyle, has reorganized existing knowledge in a useful way that may set the stage for further research.

In particular, we foresee at least three promising avenues for future studies. One is the collection and compilation of better data on a wider variety of mammals. This would allow more rigorous evaluation of predictions of scaling exponents and hypotheses about lifestyles. Another area for investigation is the relationship among phylogeny, ecology, and allometry. Sometimes ecology can exert a conservative influence, reinforcing the powerful constraints of phylogeny and allometry to retain ancestral traits. Sometimes, however, ecology can be a powerful force for change, as specialized species that have adopted new lifestyles found new lineages that diverge from their ancestors and established allometric relationships. A final area for research would involve exploring more explicitly the relationships between different components of the life history. The dependence of production rate on metabolic rate has a clear mechanistic basis. Death rates and survival may also be linked to metabolism through the effects of free radicals on aging, but this relationship may be less direct and constraining. The theory presented above shows that production and mortality rates must be complementary, but how and why this is achieved warrants further investigation. This effort will be aided by the collection and analysis of more and better data on life histories of unusual mammals, such as primates, bats, and whales.

### *Comparisons with Marsupials and Monotremes*

The above section is based on applications of metabolic scaling theory to life histories of eutherian mammals (Brown and Sibly 2006; Sibly and Brown 2007, 2009). Subsequently, data on life histories of marsupials and monotremes were compiled, analyzed, and compared with placentals (Hamilton et al. 2011).

The three lineages of mammals, placentals (eutherians), marsupials (metatherians), and monotremes (prototherians), diverged more than 150 million years ago and diversified rapidly on different landmasses. The extant members of these lineages differ conspicuously in physiology and life history. Metabolic rates of marsupials are about 33% lower and monotremes more than 50% lower than placentals. Monotremes hatch from eggs, marsupials are born at a small size and then protected in a pouch, and placentals are born at larger size and lactate until they become independent. Two common features, however, are that the mother supplies all of the nutrition to rear offspring to independence (about 33% of adult size in all mammals), and most of her energy is allocated to lactation.

Figure 9.4 shows differences among lineages and variation within lineages in life history traits. Gestation times (*A*) and mass of offspring (*B*) at birth scale very differently between lineages, increasing only a little in the larger species in marsupials, whereas masses at birth are much larger and scale steeply with size in placentals. Mass of offspring at weaning (*C*) is nearly identical. Litter size (*D*) and frequency (*E*) are highly variable in both lineages, but litter size scales more steeply and litter frequency less steeply in marsupials than placentals. Values for monotremes overlap with similar-sized marsupials.

Given the pronounced differences and variations in life histories within and between these lineages, meaningful comparisons of production should include all female investment up to the time of weaning. So we used equation 2 to calculate production rate as before but substituted weaning mass for the mass of the offspring, $M_O$. The scaling of mass-specific production rate with body size is indistinguishable across the three lineages (fig. 9.4*F*). To a first approximation all female mammals allocate a constant proportion of their energy budget to biomass of offspring. The marked differences in mammalian life history strategies appear to be alternate adaptive ways to rear offspring to independence at a rate ultimately set by the scaling of metabolic rate with body size. All mammals appear to be subject to the similar metabolic constraints on production, because they share similar body designs, vascular systems, and costs of producing new tissue. This is a striking example of how juxtaposition of theory and data on metabolic scaling can reveal unifying similarities despite substantial differences reflecting evolutionary divergence and ecological adaptation.

## Sexual Selection and Scaling of Male Traits

Whereas female mammals produce relatively small numbers of relatively large eggs over their lifetimes, males produce literally billions of miniscule sperm. Since the direct energetic and material contribution of males to offspring is negligible, fitness of males is determined primarily by how many offspring they sire. So males allocate their production to structures and behaviors that promote the competitive ability of their sperm to fertilize the limited number of eggs. The few species with substantial male parental care are exceptions to this general rule.

Ever since Darwin (1871), biologists have recognized that sexual selection favors the evolution of male characteristics that give advantages in competition for mates: body size and traits that serve either as ornaments to attract females during courtship or as weapons to fight with other males for access to females. Darwin and many subsequent authors have noted that reproductive competition and resulting sexual selection have often resulted in the evolution of "exaggerated" or "bizarre" male traits. But what do these adjectives mean, and what do they imply about sexual selection in the context of metabolic scaling theory?

Typically these traits exhibit positive allometry, increasing disproportionately with female body size. Most characteristics of organisms vary with body size either linearly or with negative allometry, so with the exponent in equation 1 equal to or less than 1. Morphological traits, such as the sizes of organs and appendages, usually scale linearly, as $M^1$. Above we have emphasized how whole-organism metabolic rate scales as $M^{3/4}$, and how this has influenced the scaling of many other traits, including many characteristics of the life history. So whole-organism production rate of females scales in direct proportion to metabolic rate, but less than linearly with body size, as $M^{3/4}$, and mass-specific production rate decreases with body size as $M^{-1/4}$. Among the very few kinds of traits that exhibit positive allometry are those that are used by males to attract and fight for females. Allocation of production to reproduction in males has resulted in larger body size and the exaggerated ornaments and weapons used to compete for mates, and these scale with an exponent substantially greater than 1. In the next sections we consider first the evolution of body size in bovids, cervids, and macropodids, and then the evolution of ornaments and weapons, focusing on the case of the antlers of deer.

### *Male Body Size: Theory and Data for Rensch's Rule*

There is a tendency across species within some taxonomic groups for the ratio of male to female body size at breeding to increase with female body size. This tendency is known as Rensch's rule (Rensch 1950; Sibly et al. 2012). Rensch's rule is not an absolute law, and there are exceptions. However, Rensch's rule does hold in three lineages of large herbivorous mammals: bovid and cervid artiodactyls and macropodid marsupials, or antelopes, deer, and kangaroos, respectively. In all three taxa, smaller species tend to live in monogamous pairs, whereas females of larger species tend to aggregate in social groups and males tend to control groups and mate with the multiple females (Jarman 1974, 1983; Geist and Bayer

1988; Croft 1989; Jarman and Coulson 1989; Geist 1998; Loison et al. 1999; Fisher and Owens 2000; Fisher et al. 2001; Croft and Eisenberg 2006; Lindenfors et al. 2007; Sibly et al. 2012). Jarman (1983) showed that in the smallest species of each family, both sexes mature rapidly, at similar body sizes, and have relatively small weapons, whereas in larger species, life span and time to sexual maturity are longer, and males grow to larger sizes than females and typically possess well-developed ornaments and weapons that are used in contests for mates. So associated with this trend of increasing polygyny are increased longevity, sexual size dimorphism, and weaponry. Here we combine models and data to reflect on the evolutionary processes that gave rise to Rensch's rule in bovids, cervids, and macropodids.

Aggregation of females in social groups provides the possibility for a single male to defend and mate with multiple females. As group size increases, more males compete to control mating access and so the intensity of sexual selection on male traits increases. Calder (1984) applied metabolic scaling to predict the scaling of female group size with body size in large herbivorous mammals. He noted that home range size of an individual female scales as $M^1$ but her metabolic rate as $M^{3/4}$, so females of larger species can aggregate and occupy overlapping home ranges. Calder predicted that the size of the group should scale as $M^{(1-3/4)} = M^{1/4}$. This prediction is shown as the faded lines in the left-hand panels of figure 9.5. In fact, group size in the bovids, cervids, and macropodids (left-hand panels of fig. 9.5) scales more steeply than Calder's prediction, and the reason for this discrepancy is not known (but see Sibly et al. 2012).

Because mammal sex ratios are close to 1:1, the number of males competing to mate with the females in a group should scale according to the empirical scaling of female group size, shown in the left-hand panels of figure 9.5. The males compete to inseminate as many females as possible, and in lineages such as bovids, cervids, and macropodids, the competition is ultimately settled largely by fighting ability. There is selection for males to grow large, because the biggest males tend to win contests, control groups, and inseminate the females. But how does a male get to be bigger than his competitors? Females stop growing when they reach sexual maturity, but males grow throughout life, so a male can get to be the biggest if he outlives the competition (Sibly et al. 2012). The time required depends on female group size and male mortality rate. The scaling of female group size can be inferred from the left-hand panels of figure 9.5. The scaling of male mortality rate is complicated and is best

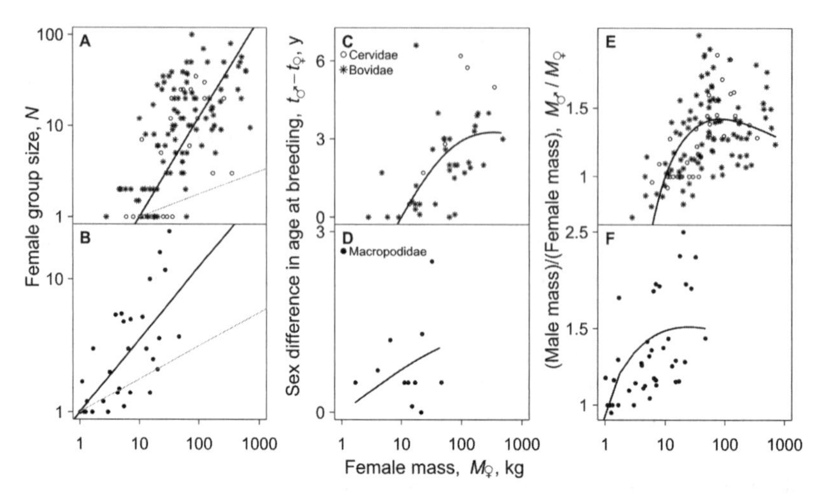


FIGURE 9.5. Data and predictions of a model of Rensch's rule. Predictions are shown as solid curved lines except in panels *A* and *B*. Data for Bovidae (*) and Cervidae (○) are shown in the upper row, for Macropodidae (●) in the lower row. Left-hand panels (*A*, *B*) test Calder's (1984) prediction, shown as faded lines, for the allometry of female group size *N*. Fitted regressions were used to obtain parameter values for model predictions in the remaining panels, *C*, *D*, *E*, and *F*. *C* and *D* test the prediction for the difference in age of first reproduction between males and females, $t♂ - t♀$. *E* and *F* test the predicted allometry of the ratio of masses of adult males and females, M♂/M♀ (Rensch's rule). From Sibly et al. (2012).

estimated empirically, because increasing size incurs increasing risk of dying. The resulting best-fit curve of predicted age at breeding, when a male is able to control a group of females, is shown in the central panels of figure 9.5.

If we know the time available for males to grow, given by the curved lines in the central panels of figure 9.5, their size at breeding can be calculated if male growth rate is known. Prior to the age of female maturation males grow at least as fast as females, but thereafter they grow more slowly. Male growth rate was estimated by data fitting, and the curves predicting final male size are shown in the right-hand panels of figure 9.5. These curves show that sexual dimorphism, that is, the ratio of body size of a successful surviving breeding male to female body size, increases with female size up to a maximum when female mass is ~100 kg and then declines. The model predicts that males should be about 50% larger than females at the peak of sexual dimorphism. However, there is considerable scatter in the data (right-hand panels of fig. 9.5), and sexual dimorphism is greater than 50% in some species. Despite the scatter in the data in all panels in figure 9.5, the fitted curves do provide reasonable descriptions of the general patterns.

A question raised by this analysis is why males grow so slowly after the age of female maturation. If males continued to grow as fast as females throughout their lives, they could mature much earlier or attain much larger sizes. Part of the explanation may be that some of the energy that could potentially be allocated to growth is expended on costly breeding behaviors, including courtship of females and aggressive contests with other males. Additionally, the increased mortality rate with increasing size and age in males suggests that the reduced growth rate may be due in part to the costs of maintaining large bodies and the ornaments and weapons used to compete for mates. These costs are probably higher in species with larger weapons, and this may explain why male growth rate is lower in bovids and cervids, where males grow horns or antlers, than in macropodids, which do not have such obvious weaponry. We next consider another effect of sexual selection on metabolic scaling, the allometry of deer antlers .

### *Allometry of Deer Antlers: Empirical Patterns*

The scaling of deer antlers has attracted the interest of such notable biologists as Julian Huxley (1932) and Stephen Jay Gould (1974; see the epigraph to this chapter). The quintessential examples of deer antlers are those of the Irish elk (*Megaloceros giganteus*), the largest deer that ever lived, which, not coincidentally, had by far the largest antlers. The antlers were so massive that authors have suggested that they contributed to the extinction of the species, which occurred at the end of the Pleistocene about 10,000 years ago.

Several authors, including Huxley (1932), Gould (1974), Geist (1966), and others (e.g., Clutton-Brock et al. 1982; Andersson 1994; Lincoln 1994; Kodric-Brown et al. 2006; Vanpé et al. 2007), have compiled data on the allometry of antlers. The exact values of the exponents vary somewhat, depending on which individuals and species are included in the analysis, but the overall result is very consistent. Antler size increases disproportionately with body size, both within species over ontogeny, and between species over phylogeny (figs. 9.6 and 9.7). Exponents are consistently substantially greater than 1, and intraspecific exponents—for ontogenetic variation—are consistently greater than interspecific

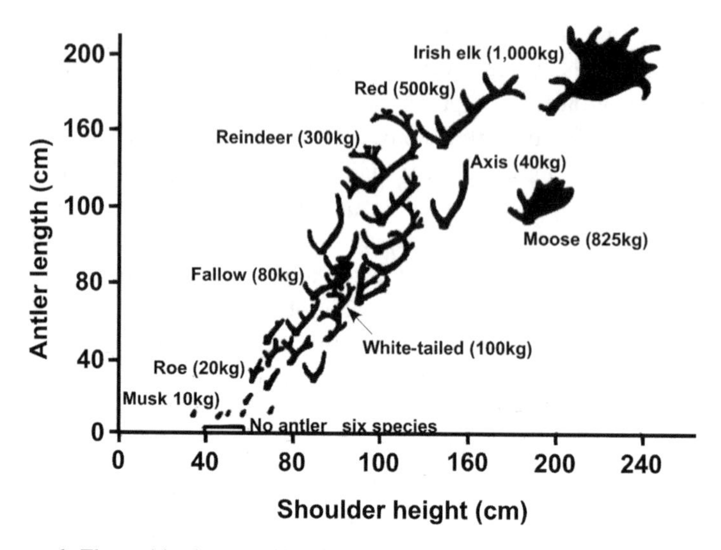


FIGURE 9.6. The positive interspecific allometry of antler length as depicted diagrammatically by Lincoln (1994). Note that here the size measurements are lengths, $l$, rather than masses, and the axes are linear rather than logarithmic. Transforming to mass, $l^3$, and fitting an allometric equation gives an exponent of 1.8.

exponents—across species of different average body size. The interspecific pattern is shown diagrammatically in the wonderful depiction of Lincoln (1994), reproduced in fig. 9.6. Although Lincoln measured antler and body size in terms of lengths and plotted his data on linear axes, reanalyzing his data in terms of mass gives an allometric exponent of 1.8. Our analysis of another dataset gives an almost identical exponent of 1.7 (fig. 9.7*A*). Obviously both exponents are much greater than 1. Intraspecific exponents are typically even higher. Within a population over ontogeny, males continue to grow and to produce proportionately larger antlers every year until they reach peak reproductive maturity (fig. 9.7*B*). Similar analyses of the horns of antelopes and other bovids give similar results (Kodric-Brown et al. 2006).

Before proceeding further it is helpful to clarify terminology. First, for convenience we shall refer to antlers as "ornaments," although we recognize full well that they have a dual function. Like many sexually selected traits, they are used both as ornaments in displays to attract females and as weapons in contests between males for access to females. Second, we restrict our analysis and discussion here to the antlers of deer, although many other traits of mammals have been elaborated by sexual selection and could be analyzed similarly. These traits include horns of bovids, courtship sounds produced by some deer, whales, and bats, and the canine teeth of hippos and some pigs, carnivores, and primates. Third, we use "intraspecific" to refer to the pattern of variation of antler size with body size over the ontogeny of an individual as it grows from juvenile to adult. We use "interspecific" to refer to variation in maximal antler size with maximal body size across different species.

Several features of deer natural history provide important background. Deer vary in body size by almost three orders of magnitude, from the so-called mouse deer, weighing 2.5 kg, to the Irish elk, weighing on the order of 1 ton. Second, as indicated above, deer have a somewhat unusual life history. Females grow to the age of first reproduction,

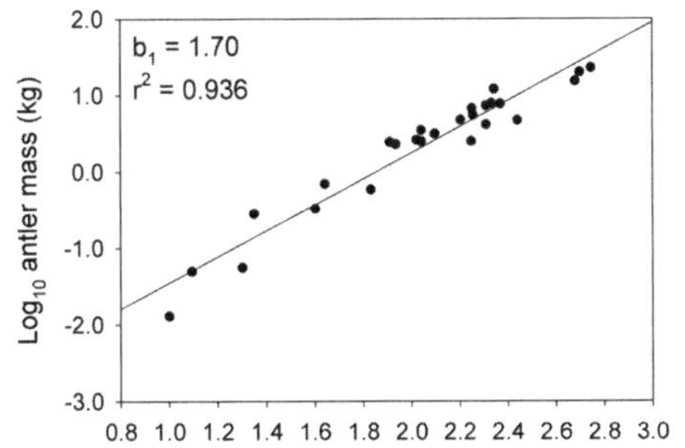


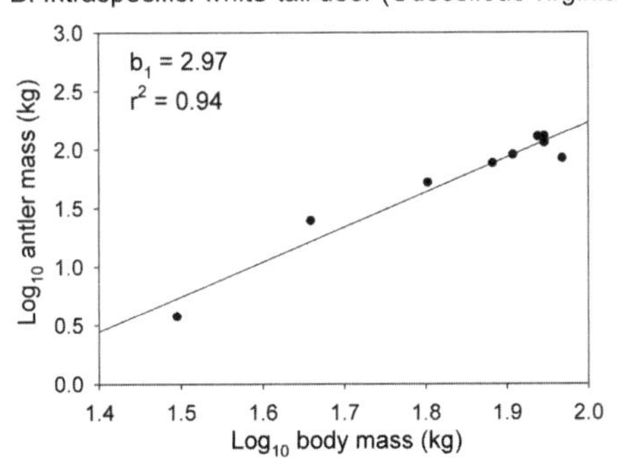


FIGURE 9.7. Positive interspecific and intraspecific allometries of deer antlers. *A*, Relationship between antler mass and body mass for 26 species of deer in the family Cervidae. Plotted are values for maximum antler mass and maximum body mass from the compilation by Geist (1998). *B*, Intraspecific ontogenetic variation in antler mass and body mass in white-tail deer. Data are from Fuller et al. (1989) for average antler and body sizes for annual age classes.

typically 1–2 years, and then maintain a nearly constant weight as they allocate production to reproduction. Males, by contrast, continue to grow, typically increasing in body size and antler size each year until reaching their prime and peak reproduction at ages varying from 1–2 in small species to 8–12 in elk and moose, as shown in fig. 9.5*C* and *E*. Antlers are unusual structures in that they are shed at the end of each breeding season and grown anew the following year. Antlers are a special form of bone. Growing them requires large quantities of minerals, notably calcium and phosphorus. For example, the antlers of a prime 479 kg elk weighed 15.3 kg (Blood and Lovaas 1966), and those of a 557 kg moose weighed 22.9 kg (Skal 1982), in each case equivalent to about half the weight of the bone in their skeletons.

### *Allometry of Deer Antlers: A Theoretical Model*

We can incorporate these features of deer natural history into a model of biomass allocation over ontogeny (fig. 9.8; see also Kodric-Brown et al. 2006). Each year a male acquires new resources and faces the "decision" of how to allocate them between somatic growth and antlers. Allocation to antlers begins at the age of sexual maturity, normally 2 years even in

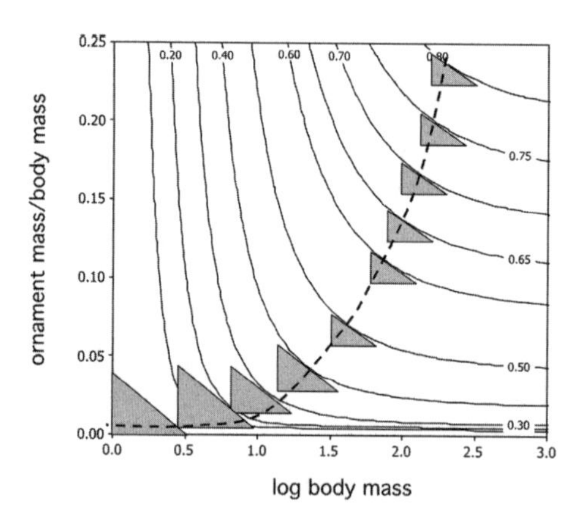


FIGURE 9.8. Diagram showing how the optimal strategy (*dashed line*) for allocation to an ornament is calculated. The options for resource allocation each year are represented as right triangles, thus resources may be allocated to increase body mass, to increase ornament size, or to both. Thin curved lines indicate achievable fitness. Note that initially the optimal strategy is to allocate all resources to increase body mass, but thereafter it pays to allocate resources increasingly to ornament. After (Kodric-Brown et al. 2006).

the largest deer species, which do not reach their prime and peak reproduction until 8–12 years. We model the allocation each year as a right triangle, whose hypotenuse reflects the trade-off between allocating to antlers and allocation to soma and hence to growth in body mass. The optimal allocation is one that maximizes fitness, which is modeled as a curvilinear function of body mass and ornament size. The optimal allocation each year is when the slope of the fitness function is just tangential to the hypotenuse of the allocation triangle. Total annual production of antler and soma was assumed to scale as the ¾ power of body mass. Unfortunately, we do not have a similar theoretical or empirical basis to derive the shape of the fitness function. So we have chosen a general mathematical form that seems appropriate until we have data or theory to provide a better characterization.

The model makes realistic predictions. The male should continue to grow in size each year, but at a decreasing rate as he allocates proportionately more resources to antlers. The result is that the optimal allocation strategy traces out a trajectory of positive allometry. The exponent predicted by this model will depend crucially on the shape of the fitness function. This in turn will depend on the ability of a male of a given size to acquire the resources required to produce antler and soma, and on the ability of the male with a given combination of antler and body size to attract females, compete with other males, and sire offspring.

The model also predicts why interspecific exponents are less than intraspecific ones, as shown in figure 9.7. If the same ontogenetic trajectories were maintained as deer speciated and evolved larger body sizes, males would soon have impossibly large antlers. Natural selection has gotten around this problem by resetting the ontogenetic trajectory as species have evolved larger size. Since larger deer species live longer and the males make a new and greater allocation to antlers every year, the largest individuals of the largest deer species end up growing proportionately larger antlers. The result is that both intraspecific and interspecific allometries are positive, but the exponents are higher for the ontogenetic relationships within species than the allometry across species (fig. 9.7*A* and *B*).

## Concluding Remarks

Putting metabolic scaling theory together with life history theory affords new insights into both the general themes and the considerable variations in mammal life histories. The most general themes are (1) the scalings of rates of production, birth, and death with female body size as $M^{-1/4}$; (2) the positive scaling of male sexual ornaments and weapons, as $M^b$, where $b > 1$. The first follows from the three simple facts that (1) whole-organism metabolic rate scales as $M^{3/4}$ (Kleiber's law); (2) metabolic rate fuels both production new biomass, and hence birth rate, and maintenance of existing biomass, and hence survival; and (3) density-dependent ecological compensation requires that over the long term birth rates equal death rates. This body of theory applies primarily to females, because it is the metabolism of the mother that supplies nearly all of the energy and materials to produce offspring. The unusual positive allometry occurs only in males, because the tiny sperm contributes negligibly to the energy and materials provided to offspring, so consequently sexual selection to compete with other males to sire offspring has favored metabolic allocation to produce larger body size and elaborate ornaments and weapons.

Variations on these general themes are substantial. These variations reflect divergence of different taxonomic or lifestyle groups from the general allometric scaling relationships for mammals as a whole. The deviations can be interpreted as variations in rates of birth, death, and production that occurred as a consequence of adaptations to different lifestyles. Marsupials differ conspicuously from placental mammals in the timing of gestation and lactation and in the allocation of maternal energy to these components of the life history. Among the placental (eutherian) mammals, marine and folivorous mammals have conspicuously high rates of production for their body size, whereas bats and primates have conspicuously low production rates and correspondingly long life spans. We interpret these patterns as reflecting the effect of abundant plant and marine food supplies in fueling high rates of metabolism and biomass production, and the effect of flying and arboreal lifestyles in reducing risk of predation. There is a strong relationship between ecology, or lifestyle, and phylogeny, or taxonomy, because the evolutionary innovations that resulted in a novel lifestyle often occurred long ago and resulted in subsequent adaptive radiations on the new theme (see Smith et al. 2004). Additionally, however, similar lifestyles and associated life histories have evolved convergently from distantly related ancestors, and sometimes they have evolved so recently as to be found only in a small clade or a single unusual species. Pervasively underlying all of the variation are common themes and seemingly near-universal scaling relationships that reflect fundamental constraints of body size on metabolic rate and of metabolism on the various component traits of the life history.

The theory and data presented here represent only the preliminary results of efforts to combine the theoretical perspectives of metabolic scaling and life history. There is still much opportunity to integrate additional insights from ecology, phylogeny, and allometry so as to understand more deeply and broadly both the general themes and the specific variations in mammal life histories. There is also much opportunity to extend these approaches to other organisms, which are likely to exhibit not only similar general themes but also distinctly different patterns of variation.

## Acknowledgments

NSF Grants DEB-0083422 to JHB and IBN-0417338 to AKB; a Royal Society Travel Grant to RMS; a training grant from the Howard Hughes Medical Institute to JHB, and NSF RCN Grant DEB 0541625 supported our collaboration. JHB and AKB thank the National Center for Ecological Analysis and Synthesis for funding and logistical support while on sabbatical there. All coauthors thank K. E. Jones for unpublished data on bats, and many individuals, but especially the UNM-SFI scaling group and E. L. Charnov, S. K. M. Ernest, M. E. Moses, M. Pagel, and F. A. Smith, for valuable discussions.

## References

Andersson, M. B. 1994. *Sexual selection*. Princeton: Princeton University Press.

Blood, D. A., and A. L. Lovaas. 1966. "Measurements and weight relationships in Manitoba elk." *Journal of Wildlife Management* 30 (1): 135–140.

Blueweiss, L., H. Fox, V. Kudzma, D. Nakashima, R. Peters, and S. Sams. 1978. "Relationships between body size and some life-history parameters." *Oecologia* 37 (2): 257–272.

Brody, S., and R. C. Proctor. 1932. "Growth and development, with special reference to domestic animals. XXIII. Relation between basal metabolism and mature body weight in different species of mammals and birds." *Missouri University Agricultural Experiment Station Research Bulletin* 166:89–101.

Brown, J. H., J. F. Gillooly, A. P. Allen, V. M. Savage, and G. B. West. 2004. "Toward a metabolic theory of ecology." *Ecology* 85 (7): 1771–1789.

Brown, J. H., and R. M. Sibly. 2006. "Life-history evolution under a production constraint." *Proceedings of the National Academy of Sciences of the United States of America* 103:17595–17599.

Calder, W. A. 1984. *Size, function, and life history.* Cambridge, MA: Harvard University Press.

Charnov, E. L. 1991. "Evolution of life-history variation among female mammals." *Proceedings of the National Academy of Sciences of the United States of America* 88:1134–1137.

——. 1993. *Life history invariants : Some explorations of symmetry in evolutionary ecology.* Oxford Series in Ecology and Evolution. Oxford: Oxford University Press.

Charnov, E. L., R. Warne, and M. Moses. 2007. "Lifetime reproductive effort." *American Naturalist* 170 (6):E129–E142. doi: 10.1086/522840.

Clutton-Brock, T. H., F. E. Guinness, and S. D. Albon. 1982. *Red deer: Behavior and ecology of two sexes.* Chicago: University of Chicago Press.

Croft, D. B. 1989. "Social organization of the Macropoda." In *Kangaroos, wallabies and rat-kangaroos*, edited by G. Grigg, P. Jarman, and I. Hume, 505–525. Chipping Norton, NSW: Surrey Beatty and Sons.

Croft, D. B., and J. F. Eisenberg. 2006. "Behaviour." In *Marsupials*, edited by P. J. Armati, C. R. Dickman, and I. D. Hume, 229–298. Cambridge: Cambridge University Press.

Darwin, C. 1871. *The descent of man and selection in relation to sex.* Vol. 1. London: John Murray.

Ernest, S. K. M. 2003. "Life history characteristics of placental non-volant mammals." *Ecology* 84:3402.

Fisher, D. O., and I. P. F. Owens. 2000. "Female home range size and the evolution of social organization in macropod marsupials. ." *Journal of Animal Ecology* 69:1083–1098.

Fisher, D. O., I. P. F. Owens, and C. N. Johnson. 2001. "The ecological basis of life history variation in marsupials." *Ecology* 82:3531–3540.

Fuller, T. K., R. M. Pace III, J. A. Markl, and P. L. Coy. 1989. "Morphometrics of white-tailed deer in north-central Minnesota." *Journal of Mammalogy* 70:184–188.

Geist, V. 1966. "Evolution of horn-like organs." *Behaviour* 27:175–214.

——. 1998. *Deer of the world: Their evolution, behaviour, and ecology.* Mechanicsburg, PA: Stackpole Books.

Geist, V., and M. Bayer. 1988. "Sexual dimorphism in the Cervidae and its relation to habitat." *Journal of Zoology (London)* 214:45–53.

Gould, S. J. 1974. "The origin and function of 'bizarre' structures: Antler size and skull size in the 'Irish elk,' *Megaloceros giganteus*." *Evolution* 28 (2): 191–220.

Hamilton, M. J., A. D. Davidson, R. M. Sibly, and J. H. Brown. 2011. "Universal scaling of production rates across mammalian lineages." *Proceedings of the Royal Society B: Biological Sciences* 278:560–566.

Harvey, P. H., and T. H. Clutton-Brock. 1985. "Life history variation in primates." *Evolution* 39:559–581.

Harvey, P. H., and S. Nee. 1991. "How to live like a mammal." *Nature* 350:23–24.

Harvey, P. H., M. D. Pagel, and J. A. Rees. 1991. "Mammalian metabolism and life histories." *American Naturalist* 137 (4): 556–566.

Harvey, P. K. , and P. M. Bennett. 1983. "Brain size, energetics, ecology, and life history patterns." *Nature* 306:314–315.

Huxley, J. S. 1932. *The problems of relative growth*. London: Methuen.

Jarman, P. J. 1974. "The social organization of antelope in relation to their ecology." *Behaviour* 48:215–267.

——. 1983. "Mating system and sexual dimorphism in large, terrestrial, mammalian herbivores." *Biological Reviews* 58:485–520.

Jarman, P. J., and G. Coulson. 1989. "Dynamics and adaptiveness of grouping in macropods." In *Kangaroos, wallabies and rat-kangaroos*, edited by G. Grigg, P. Jarman, and I. Hume. Chipping Norton, NSW: Surrey Beatty and Sons.

Kleiber, M. 1932. "Body size and metabolism." *Hilgardia* 6:315–351.

Kodric-Brown, A., R. M. Sibly, and J. H. Brown. 2006. "The allometry of ornaments and weapons "*Proceedings of the National Academy of Sciences of the United States of America* 103:8733–8738.

Lincoln, G. A. 1994. "Teeth, horns and antlers: The weapons of sex." In *The difference between the sexes*, edited by R. V. Short and E. Balban, 131–158. Cambridge: Cambridge University Press.

Lindenfors, P., J. L. Gittleman, and K. E. Jones. 2007. "Sexual size dimorphism in mammals." In *Sex, size and gender roles*, edited by D. J. Fairburn, W. U. Blanckenhorn, and T. Szekely, 16–26. Oxford: Oxford University Press.

Loison, A., J. M. Gaillard, C. Pelabon, and N. G. Yoccoz. 1999. "What factors shape sexual size dimorphism in ungulates?" *Evolutionary Ecology Research* 1:611–633.

McMahon, T. A., and J. T. Bonner. 1983. *On size and life*. New York: Scientific American Books.

McNab, B. K. 1986. "The influence of food-habits on the energetics of Eutherian mammals." *Ecological Monographs* 56 (1): 1–19.

——. 2002. *The physiological ecology of vertebrates: A view from energetics*. Ithaca: Cornell University Press.

Meiri, S., J. H. Brown, and R. M. Sibly. 2011. "The ecology of lizard reproductive output." *Global Ecology and Biogeography* 21:592–602.

Peters, R. H. 1983. *The ecological implications of body size*. Cambridge: Cambridge University Press.

Promislow, D. E. L., and P. H. Harvey. 1990. "Living fast and dying young: A comparative analysis of life-history variation among mammals." *Journal of Zoology* 220:417–437.

Purvis, A., and P. H. Harvey. 1997. "The right size for a mammal." *Nature* 386:332–333.

Read, A. F., and P. H. Harvey. 1989. "Life-history differences among the Eutherian radiations." *Journal of Zoology* 219:329–353.

Rensch, B. 1950. "Die Abhängigkeit der relativen sexualdifferenz von der Kőrpergrőße." *Bonner Zoologische Beiträge* 1:58–69.

Ross, C. 1998. "Primate life histories." *Evolutionary Anthropology* 6:54–63.

Schmidt-Nielsen, K. 1984. *Scaling, why is animal size so important?* Cambridge: Cambridge University Press.

Sibly, R. M., and J. H. Brown. 2007. "Effects of body size and lifestyle on evolution of mammal life histories." *Proceedings of the National Academy of Sciences of the United States of America* 104:17707–17712.

——. 2009. "Mammal reproductive strategies driven by offspring mortality-size relationships." *American Naturalist* 176:E185–E199.

Sibly, R. M., W. Zuo, A. Kodric-Brown, and J. H. Brown. 2012. "Rensch's rule in large herbivorous mammals derived from metabolic scaling." *American Naturalist* 179:169–177.

Skal, O. 1982. *Jagdparadies Alaska*. 2nd ed. Graz: L. Stocker.

Smith, F. A., J. H. Brown, J. P. Haskell, S. K. Lyons, J. Alroy, E. L. Charnov, T. Dayan, B. J. Enquist, S. K. M. Ernest, E. A. Hadly, K. E. Jones, D. M. Kaufman, P. A. Marquet, B. A. Maurer, K. J. Niklas, W. P. Porter, B. Tiffney, and M. R. Willig. 2004. "Similarity of mammalian body size across the taxonomic hierarchy and across space and time." *American Naturalist* 163 (5): 672–691.

Thompson, D. W. 1917. *On growth and form*. Cambridge: Cambridge University Press.

Vanpé, C., J. M. Gaillard, P. Kjellander, A. Mysterud, P. Magnien, D. Delorme, G. Van Laere, F. Klein, O. Liberg, and A. J. M. Hewison. 2007. "Antler size provides an honest signal of male phenotypic quality in roe deer." *American Naturalist* 169 (4): 481–493.

West, G. B., J. H. Brown, and B. J. Enquist. 1997. "A general model for the origin of allometric scaling laws in biology." *Science* 276 (5309): 122–126.

——. 1999. "The fourth dimension of life: Fractal geometry and allometric scaling of organisms." *Science* 284:1677–1679.

CONCLUSION

# The Way Forward

Felisa A. Smith and S. Kathleen Lyons

A number of common themes run throughout this volume. First, one begins to appreciate just how much variation there is in body size among most groups of animals. For the most part, a clade is not a homogenous collection of similar bauplans. Even among volant taxa, such as birds and bats, substantial variation in body mass exists despite the considerable constraints imposed by flight (chaps. 3, 4, 8). Moreover, body size variation is clearly important in terms of community and ecosystem dynamics and structure (e.g., chaps. 6–9); how animals interact with their environment is strongly mediated by their body mass. Thus, the size of an organism is of great physiological and ecological significance (Peters 1983).

Second, there are remarkably consistent patterns in the body size distributions of some taxa across spatial and temporal gradients that may well reflect how organisms acquire and allocate energy. For example, right-skewed body size distributions seem to be quite common among diverse clades (chaps. 1–5). However, these regular patterns vary between groups in ways that we do not yet understand. As Maurer and Marquet (chap. 7) note, numerous processes have been invoked to explain the minimum, maximum, mode, and shape of body mass distributions. Which of these processes are universal? Which are unique to particular taxa or geological histories (e.g., response to glaciation; chap. 2)? Are size frequency distributions similarly skewed to the right because organisms contributing to these distributions share similar trophic or life history traits? To some extent, our inability to develop universal explanations may reflect a lack of fundamental information. For example, as Gaston and Chown (chap. 1) note, it is likely that only 9%–24% of extant insect species have been described thus far. How robust are generalizations for insects based on such a restricted sampling of extant biodiversity?

Third, temperature is important. Many taxa demonstrate a clear response to underlying environmental gradients in temperature (chaps. 1, 2, 5). This ecogeographic pattern is known as Bergmann's rule: the principle that within a broadly distributed genus, species of larger size are found in colder environments, and species of smaller size are found in warmer areas (Rensch 1938, 1950; Mayr 1956). While Bergmann's rule holds for the majority (62%–83%) of vertebrates examined to date (Millien 2006), it is not clear how robustly other clades follow this pattern. Evidence within our volume is mixed; some insect groups do conform (fig. 1.4), but land snails apparently do not (fig. 2.1). Here, the role of evolutionary history may be particularly important, as much of the current diversity of snails may be related to the most recent glaciation events of the Pleistocene. Interestingly, dispersal may be implicated for the lack of a Bergmann-like cline, not only in snails (chap. 2), but in bats as well (chap. 4). Clearly, understanding the influence of temperature on organism body size is particularly important in this era of anthropogenic climate change. While considerable effort has gone into predicting how species will respond to the large anthropogenic climate shifts expected over the next few centuries, most studies focus on changes in the distribution and abundance of species (reviewed in Parmesan and Yohe 2003; Root et al. 2003; Ackerly et al. 2010). They tend to completely discount the possibility of in situ adaptation. Yet Bergmann clines over space and time demonstrate the ability of species to adapt to fluctuating abiotic conditions and highlight the strong selection imposed by the environment (Smith et al. 1995; Smith and Betancourt 2006). As the contributions in our volume demonstrate, changing environmental conditions may well result in shifts in the body size of organisms and communities, leading to wholesale alterations in energy use.

The observant reader will quickly notice several gaps in this compilation. First, our focus here has been exclusively on animals. A logical question is whether body size patterns across widely divergent groups display similar features. A qualified answer to this question is yes. At least for plants, where data compilations exist, strikingly regular patterns of "body size" exist. In the last decade, Brian Enquist, Karl Niklas, and others have made considerable progress in connecting individual attributes of plants with macroscopic patterns (Niklas 1994). For example, plants show the same scaling of population density and body mass seen with animals (Niklas and Enquist 2001; Enquist and Niklas 2002). And vascular plants show at least as great a range of body mass variation as other taxa; approximately twelve orders of magnitude in mass (Enquist et al. 1999; Niklas and Enquist 2001). Much of this morphological variation is significantly related to structural complexity and life history diversity (Niklas 1994; Westoby 1998; Enquist et al. 1999; Enquist and Niklas 2002; Niklas and Enquist 2002a, 2002b; Westoby et al. 2002). Clearly, comparing patterns of body size, abundance, and/or distribution among diverse groups is likely to be extremely fruitful and an area where only a few workers have yet explored (e.g., Ernest et al. 2003; Hechinger et al. 2011).

But, even within animals, there are important groups lacking from our volume. With some exceptions, such as that of marine mollusks, not considered here because there is an entire volume on marine macroecology (Witman and Roy 2009), this largely reflects a lack of comprehensive data. For example, despite the considerable interest demonstrated by the public and others in dinosaurs, only a handful of studies have begun to try to use a macroscopic perspective to examine the distribution or abundance of species (Carrano 2006), although such efforts are ongoing (table C.1). Potential authors we contacted felt these data were too poor at the present time to try to synthesize patterns. In other instances, excellent data exist, but are focused on specific taxa limiting their utility for evaluating synoptic patterns. Research on fish, for example, has tended to focus on a few abundant species of commercial significance (Rice et al. 1991; Christensen 1995; Jennings et al. 2001). Indeed, the effects of overharvesting are well known to influence the body size distribution of remaining stocks (Law and Rowell 1993; Rowell 1993; Stokes and Blythe 1993; Law and Stokes 2005). However, we lack a comprehensive understanding of the overall body size distribution of animals in the world's oceans.

Second, despite the great strides made in molecular techniques over the last decade, for most groups a detailed species-level phylogeny is still lacking. This hampers our ability to examine the influence of evolutionary history on the body size patterns of animals. Just what are the relative contributions of phylogenetic autocorrelation, environmental factors, and architectural limitations? The authors of most chapters did attempt to incorporate phylogenies into analyses, but they often used taxonomy as a proxy for a true phylogenic relationship. Clearly, as more

TABLE C.1 **Synthetic Databases That Currently Exist That Contain Body Size Estimates of Various Taxa**

| Taxa | Database | Source and Reference | Spatial Extent | Temporal Extent | Description |
|---|---|---|---|---|---|
| All | PBDB | Paleobiology Database, http://paleodb.org; organized and operated by an international group of paleobiological researchers led by J. Alroy | Global | Phanerozoic | Collection-based occurrence, taxonomic, and some measurement data for animals and plants; web-based software for statistical analysis |
| Coleoptera (beetles) | | Buckland and Buckland 2006 | Global | Pleistocene; 2 million years ago to present | Species-level habitat and distribution data; includes tools for climate and environmental reconstruction |
| Fish | Fishbase2004 | http://www.fishbase.org/search.php; supported by a consortium of nine research institutions | Global | Modern | Species-level biological and habitat data; includes mass and/or size, habitat, climate, description, biology and International Union for Conservation of Nature status |
| Lizards | | Meiri 2008 | Global | Modern | Species-level estimates of snout-vent length (mm) and mass (g) |
| Mammals | MOM v3.0 | http://www.esapubs.org/archive/ecol/E084/094/; latest version: http://biology.unm.edu/fasmith/; Smith et al. 2003 | Global | Late Quaternary | Species-level biological and habitat data; includes body mass, distribution, trophic characterization |
| Mammals | Pantheria | http://www.esajournals.org/doi/abs/10.1890/08–1494.1; Jones et al. 2009 | Global | Historic | Species-level biological and habitat data; includes body mass, life history, distribution |
| Mammals | Mammal_lifehistories_v2.0 | http://www.esapubs.org/archive/ecol/E084/093/; Ernest 2003 | Global | Modern | Mammal life history characteristics |

| | | | | | |
|---|---|---|---|---|---|
| Mammals | Mammoth v1.0 | latest version: http://biology.unm.edu/fasmith/ | Global | Cenozoic | Maximum body mass of each mammalian order by subepoch, on each continent |
| Mammals | NOW | http://www.helsinki.fi/science/now/; M. Fortelius (coordinator), Neogene of the Old World Database of Fossil Mammals (NOW), University of Helsinki, http://www.helsinki.fi/science/now/ | | | Neogene of the Old World; Eurasian Miocene to Pleistocene land mammal taxa and localities, with emphasis on the European Miocene and Pliocene |
| Mammals | FAUNMAP | http://www.museum.state.il.us/research/faunmap/query/ | North America | Pliocene to present (5 million years ago to present) | Mammal fossils |
| Mammals | MIOMAP | http://www.ucmp.berkeley.edu/miomap/ | North America | Oligocene to Pliocene (30 million to 5 million years ago) | Mammal fossils |
| Snakes | | http://www.auburn.edu/academic/science_math/cosam/collections/reptiles_amphibians/projects/index.htm; Boback and Guyer 2003 | Regional | Modern | Body sizes (log10 [maximum total length]) of the largest and smallest species of snake in island assemblages and island area; body sizes (maximum snout-vent length) from island and mainland populations for 30 species of snakes |
| Snakes | | http://www.auburn.edu/academic/science_math/cosam/collections/reptiles_amphibians/projects/index.htm; Boback and Guyer 2008 | ??? | Modern | Body sizes (maximum total length, except for turtles, which are maximum carapace length) of vertebrates |

detailed phylogenies become available, it will become possible to incorporate such information.

Third, a historical perspective is missing for most groups despite the best efforts of the chapters' authors to incorporate evolutionary history. Yet, a major question is whether the emergent statistical patterns observed over space also exist across time. And, if so, just how consistent have they been? Consistency across space and time would strongly suggest universal drivers. Moreover, the role of size in extinction and/or speciation is crucial to understanding current diversity and body size distributions. Several studies have examined the influence of body size in influencing species richness (e.g., Dial and Marzluff 1988; Gittleman and Purvis 1998; Orme et al. 2002), but more comprehensive tests across different lineages are required.

The lack of a comprehensive historical perspective for mammals was the reason we put together our Research Coordination Network (IMPPS; http://biology.unm.edu/impps_rcn/). Over the past five years, this group has focused on assembling body mass estimates for fossil lineages to allow comparison of body mass patterns over space with those over time. Recently, we published several papers examining the trajectory of mammalian body mass on the various continents over evolutionary time (Smith et al. 2010; Evans et al. 2012). Our results confirm much of what we know about mammalian body mass over geographic space (chap. 5). The evolutionary trajectory was quite similar on all continents across the last 100 million years, even within clades. It appears that the upper limit on mammalian body mass is related to both temperature and area of the continent (Smith et al. 2010). Patterns within orders are extremely similar on all continents, although there are interesting differences that relate to the evolutionary history of taxa. Our work in this area is ongoing.

Collecting these data was a major effort of our group. Consider that assembling our database Mammoth (v 1.0) took a core group of 10–12 individuals with different expertise on mammalian clades about three years; the equivalent of a single researcher working for nearly thirty years! Clearly, this is only feasible with a large collaborative group. But such large synthetic databases are essential if we are to detect patterns and processes not discernible at smaller spatial or temporal scales. Unfortunately, for projects such as ours that can take so long come to fruition, there is still little institutional funding. We were fortunate to secure funding through the National Science Foundation for our efforts, but such grants are quite difficult to obtain. Thus, it is no wonder that comprehensive body size data for other terrestrial animal groups over deep time are still largely lacking.

As an aside, a major exception to the paucity of deep-time data is the marine invertebrate record. Here, extensive work has led to a reasonably complete picture of species occurrence (and in some instances, size) over the Phanerozoic (e.g., Sepkoski et al. 1981; Labandeira and Sepkoski 1993; Alroy et al. 2001, 2008). This compilation was jump-started by Jack Sepkoski, who began synthesizing diversity patterns over the Phanerozoic (e.g., Sepkoski et al. 1981; Sepkoski 1988, 1993). Since then, the Paleobiology Database (http://paleodb.org/cgi-bin/bridge.pl) has been the focus of intensive data collection by a multidisciplinary, multi-institutional, and international group of paleobiological researchers. Ultimately, the aim is to provide global data for marine and terrestrial animals and plants of any geological age. However, even this august group has had problems securing the appropriate funding.

As we move forward, it is clear that progress will depend on the continued development of large databases relating important attributes of organisms with body size and abundance. We have discussed the hurdles—often financial—in developing such databases. But, other issues remain. For example, different disciplines estimate "body size" in very different ways. While mammalogists routinely use mass (chaps. 5, 8), length (e.g., chaps. 1, 4), height (chap. 2), diameter (plants; Niklas 1994), and "biovolume" (aquatic groups; McClain et al. 2009) are also surrogates for an organism's size. Paleontologists may use surface area (e.g., Jablonski 1997). Converting these metrics into a standard format is essential for synthetic studies (Payne et al. 2009).

Another issue that plagues the construction of large-scale databases is that of scientific attribution. Increasingly, journals relegate methods to online supplementary material, which are generally not indexed by electronic search engines (e.g., PubMed, Web of Science, Scopus, Google Scholar). This means that primary data, often from single investigators, are not credited in large databases, undervaluing their importance and lessening the impetus for workers to "share" data (Payne et al. 2012). Moreover, this also leads to a systematic undervaluation of the subdisciplines that disproportionately provide foundational data, such as taxonomy, systematics, and natural history. Yet large data compilations are dependent on the availability of large numbers of primary studies, which must be properly acknowledged.

Finally, we lack good methods for statistically analyzing large-scale patterns. The authors of the various chapters have used different tool kits, which provide slightly different perspectives but are not always comparable. The problem of appropriate statistical methods is one that pervades macroecological approaches, which are commonly used to examine the patterns and underlying causal mechanisms of body size (Smith et al. 2008). The use of nonexperimental or "natural" data and the broad geographic, taxonomic, and/or temporal scales mean that many modern statistical methods, which have been developed for traditional experimentally based science, are inappropriate. This is an area that requires further development.

While our volume does not provide answers to all the intriguing questions raised by the past few decades of research on body size, it does highlight many of these. Clearly, there are a number of profoundly important questions that remain unaddressed, including: (1) Why are distributions of mammals on the various continents so similar despite different taxonomies, geological conditions, and environments? (2) What limits the upper and lower size of organisms? (3) Does body size influence origination or extinction rates? Do large animals evolve "slower," or are they more prone to extinction because of lowered productivity and densities, or less prone because of larger geographic ranges? and (4) Are there "rules" to body size evolution that are universal across organisms, modes of life, and/or life histories? One promising avenue may be the application of metabolic ecology to broad-scale patterns of body size. Metabolic ecology suggests that when corrections are made for body size, virtually all living things convert resources at a similar rate because of the geometry of transport systems (West et al. 1997, 1999). Clearly, the study of body size, the factors influencing it, and its attendant properties, are likely to be fruitful areas for further research for years to come.

## References

Ackerly, D. D., S. R. Loarie, W. K. Cornwell, S. B. Weiss, H. Hamilton, R. Branciforte, and N. J. B. Kraft. 2010. "The geography of climate change: Implications for conservation biogeography." Diversity and Distributions 16 (3): 476–487. doi: 10.1111/j.1472-4642.2010.00654.x.

Alroy, J., M. Aberhan, D. J. Bottjer, M. Foote, F. T. Fursich, P. J. Harries, A. J. W. Hendy, S. M. Holland, L. C. Ivany, W. Kiessling, M. A. Kosnik, C. R.

Marshall, A. J. McGowan, A. I. Miller, T. D. Olszewski, M. E. Patzkowsky, S. E. Peters, L. Villier, P. J. Wagner, N. Bonuso, P. S. Borkow, B. Brenneis, M. E. Clapham, L. M. Fall, C. A. Ferguson, V. L. Hanson, A. Z. Krug, K. M. Layou, E. H. Leckey, S. Nurnberg, C. M. Powers, J. A. Sessa, C. Simpson, A. Tomasovych, and C. C. Visaggi. 2008. "Phanerozoic trends in the global diversity of marine invertebrates." Science 321 (5885): 97–100.

Alroy, J., C. R. Marshall, R. K. Bambach, K. Bezusko, M. Foote, F. T. Fursich, T. A. Hansen, S. M. Holland, L. C. Ivany, D. Jablonski, D. K. Jacobs, D. C. Jones, M. A. Kosnik, S. Lidgard, S. Low, A. I. Miller, P. M. Novack-Gottshall, T. D. Olszewski, M. E. Patzkowsky, D. M. Raup, K. Roy, J. J. Sepkoski, M. G. Sommers, P. J. Wagner, and A. Webber. 2001. "Effects of sampling standardization on estimates of Phanerozoic marine diversification." Proceedings of the National Academy of Sciences of the United States of America 98 (11): 6261–6266.

Boback, S. M., and C. Guyer. 2003. "Empirical evidence for an optimal body size in snakes." Evolution 57:345–351.

——. 2008. "A test of reproductive power in snakes." Ecology 89 (5): 1428–1435.

Buckland, P. I., and P. C. Buckland. 2006. BugsCEP Coleopteran Ecology Package. IGBP PAGES/World Data Center for Paleoclimatology Data Contribution Series 2006-116. Boulder, CO: NOAA/NCDC Paleoclimatology Program. http://www.ncdc.noaa.gov/paleo/insect.html.

Carrano, M. T. 2006. "Body-size evolution in the Dinosauria." In Amniote paleobiology: Perspectives on the evolution of mammals birds and reptiles, edited by M. T. Carrano, R. W. Blob, T. J. Gaudin, and J. R. Wible, 225–268. Chicago: University of Chicago Press.

Christensen, V. 1995. "A model of trophic interactions in the North Sea in 1981, the year of the stomach." Dana 11:1–28.

Dial, K. P., and J. M. Marzluff. 1988. "Are the smallest organisms the most diverse?" Ecology 69:1620–1624.

Enquist, B. J., and K. J. Niklas. 2002. "Global allocation rules for patterns of biomass partitioning in seed plants." Science 295 (5559): 1517–1520.

Enquist, B. J., G. B. West, E. L. Charnov, and J. H. Brown. 1999. "Allometric scaling of production and life-history variation in vascular plants." Nature 401 (6756): 907–911.

Ernest, S. K. M. 2003. "Life history characteristics of placental nonvolant mammals." Ecology 84:3402.

Ernest, S. K. M., B. J. Enquist, J. H. Brown, E. L. Charnov, J. F. Gillooly, V. Savage, E. P. White, F. A. Smith, E. A. Hadly, J. P. Haskell, S. K. Lyons, B. A. Maurer, K. J. Niklas, and B. Tiffney. 2003. "Thermodynamic and metabolic effects on the scaling of production and population energy use." Ecology Letters 6 (11): 990–995.

Evans, A. R., D. Jones, A. G. Boyer, J. H. Brown, D. P. Costa, S. K. M. Ernest, E. M. G. Fitzgerald, M. Fortelius, J. L. Gittleman, M. J. Hamilton, L. E. Harding, K. Lintulaakso, S. K. Lyons, J. G. Okie, J. J. Saarinen, R. M. Sibly, F. A. Smith, P. R. Stephens, J. M. Theodor, and M. D. Uhen. 2012. "The maximum rate of mammal evolution." Proceedings of the National Academy of Sciences of the United States of America 109:4187–4190.

Gittleman, J. L., and A. Purvis. 1998. "Body size and species-richness in carnivores and primates." Proceedings of the Royal Society B: Biological Sciences 265:113–119.

Hechinger, R. F., K. D. Lafferty, A. P. Dodson, J. H. Brown, and A. M. Kuris. 2011. "A common scaling rule for abundance, energetics, and production of parasitic and free-living species." Science 333:445–448.

Jablonski, D. 1997. "Body-size evolution in Cretaceous molluscs and the status of Cope's rule." Nature 385 (6613): 250–252.

Jennings, S., J. K. Pinnegar, N. V. C. Polunin, and T. W. Boon. 2001. "Weak cross-species relationships between body size and trophic level belie powerful size-based trophic structuring in fish communities." Journal of Animal Ecology 70 (6): 934–944.

Jones, K. E., J. Bielby, M. Cardillo, S. A. Fritz, J. O'Dell, C. D. L. Orme, K. Safi, W. Sechrest, E. H. Boakes, C. Carbone, C. Connolly, M. L. J. Cutts, J. K. Foster, R. Grenyer, M. Habib, C. A. Plaster, S. A. Price, E. A. Rigby, J. Rist, A. Teacher, O. R. P. Bininda-Emonds, J. L. Gittleman, G. M. Mace, and A. Purvis. 2009. "PanTHERIA: A species-level database of life-history, ecology and geography of extant and recently extinct mammals." Ecology 90:2648.

Labandeira, C. C., and J. J. Sepkoski. 1993. "Insect diversity in the fossil record." Science 261 (5119): 310–315.

Law, R., and C. A. Rowell. 1993. "Cohort structured populations, selection responses, and exploitation of the North Sea cod." In The exploitation of evolving resources, edited by T. K. Stokes, J. M. McGlade, and R. Law, 155–173. Berlin: Springer-Verlag.

Law, R., and T. K. Stokes. 2005. "Evolutionary impacts of fishing on target populations." In Marine Conservation Biology: The Science of Maintaining the Sea's Biodiversity, edited by L. Crowder and E. Norse, 232–246. New York: Island Press.

Mayr, E. 1956. "Geographical character gradients and climatic adaptation." Evolution (10): 105–108.

McClain, C. R., M. A. Rex, and R. Etter. 2009. "Patterns in deep-sea macroecology." In Marine macroecology, edited by J. D. Witman and K. Roy, 65–100. Chicago: University of Chicago Press.

Meiri, S. 2008. "Evolution and ecology of lizard body sizes." Global Ecology and Biogeography 17:724–734.

Millien, V. 2006. "Morphological evolution is accelerated among island mammals." PLoS Biology 4:1863–1868.

Niklas, K. J. 1994. Plant allometry: The scaling of form and process. Chicago: University of Chicago Press.

Niklas, K. J., and B. J. Enquist. 2001. "Invariant scaling relationships for interspecific plant biomass production rates and body size." Proceedings of the National Academy of Sciences of the United States of America 98 (5): 2922–2927.

——. 2002a. "Canonical rules for plant organ biomass partitioning and annual allocation." American Journal of Botany 89 (5): 812–819.

——. 2002b. "On the vegetative biomass partitioning of seed plant leaves, stems, and roots." American Naturalist 159 (5): 482–497.

Orme, C. D. L., N. J. B. Isaac, and A. Purvis. 2002. "Are most species small? Not within species-level phylogenies." Proceedings of the Royal Society B: Biological Sciences 269 (1497): 1279–1287. doi: 10.1098/rspb.2002.2003.

Parmesan, C., and G. Yohe. 2003. "A globally coherent fingerprint of climate change impacts across natural systems." Nature 421 (6918): 37–42.

Payne, J. L., A. G. Boyer, J. H. Brown, S. Finnegan, M. Kowalewski, R. A. Krause, S. K. Lyons, C. R. McClain, D. W. McShea, P. M. Novack-Gottshall, F. A. Smith, J. A. Stempien, and S. C. Wang. 2009. "Two-phase increase in the maximum size of life over 3.5 billion years reflects biological innovation and environmental opportunity." Proceedings of the National Academy of Sciences of the United States of America 106 (1): 24–27.

Payne, J. L., F. A. Smith, M. Kowalewski, R. A. Krause, Jr., A. G. Boyer, C. R. McClain, S. Finnegan, and P. M. Novack-Gottshall. 2012. "A lack of attribution: The need to reform citation and indexing practices in the sciences." Taxon 61:1349–1354.

Peters, R. H. 1983. The ecological implications of body size. Cambridge: Cambridge University Press.

Rensch, B. 1938. "Some problems of geographical variation and species-formation." Proceedings of the Linnean Society of London 150:275–285.

——. 1950. "Die Abhängigkeit der relativen Sexualdifferenz von der Kőrpergrőße." Bonner Zoologische Beiträge 1:58–69.

Rice, J. C., N. Daan, J. G. Pope, and H. Gislason. 1991. "The stability of estimates of suitabilities in MSVPA over four years of data from predator stomachs." ICES Marine Science Symposia 193:34–45.

Root, T. L., J. T. Price, K. R. Hall, S. H. Schneider, C. Rosenzweig, and J. A. Pounds. 2003. "Fingerprints of global warming on wild animals and plants." Nature 421 (6918): 57–60.

Rowell, C. A. 1993. "The effects of fishing on the timing of maturity in North Sea cod (Gadus morhua L.)." In The exploitation of evolving resources, edited by T. K. Stokes, J. M. McGlade, and R. Law, 44–61. Berlin: Springer-Verlag.

Sepkoski, J. J., R. K. Bambach, D. M. Raup, and J. W. Valentine. 1981. "Phanerozoic marine diversity and the fossil record." Nature 293, 435–437.

Sepkoski, J. J., Jr. 1988. "Alpha, beta, or gamma: Where does all the diversity go?" Paleobiology 14 (3): 221–234.

——. 1993. "Ten years in the library: New data confirm paleontological patterns." Paleobiology 19 (1): 43–51.

Smith, F. A., and J. L. Betancourt. 2006. "Predicting woodrat (*Neotoma*) responses to anthropogenic warming from studies of the palaeomidden record." Journal of Biogeography 33:2061–2076.

Smith, F. A., J. L. Betancourt, and J. H. Brown. 1995. "Evolution of body-size in the woodrat over the past 25,000 years of climate-change." Science 270 (5244): 2012–2014.

Smith, F. A., A. G. Boyer, J. H. Brown, D. P. Costa, T. Dayan, S. K. M. Ernest, A. R. Evans, M. Fortelius, J. L. Gittleman, M. J. Hamilton, L. E. Harding, K. Lintulaakso, S. K. Lyons, C. McCain, J. G. Okie, J. J. Saarinen, R. M. Sibly, P. R. Stephens, J. Theodor, and M. D. Uhen. 2010. "The evolution of maximum body size of terrestrial mammals." Science 330 (6008): 1216–1219. doi: 10.1126/science.1194830.

Smith, F. A., S. K. Lyons, S. K. M. Ernest, and J. H. Brown. 2008. "Macroecology: More than the division of food and space among species on continents." Progress in Physical Geography 32 (2): 115–138.

Smith, F. A., S. K. Lyons, S. K. M. Ernest, K. E. Jones, D. M. Kaufman, T. Dayan, P. A. Marquet, J. H. Brown, and J. P. Haskell. 2003. "Body mass of late Quaternary mammals." Ecology 84:3402.

Stokes, T. K., and S. P. Blythe. 1993. "Size selective harvesting and age at maturity. 2. Real populations and management options." In The exploitation of evolving resources, edited by T. K. Stokes, J. M. McGlade, and R. Law, 232–247. Berlin: Springer-Verlag.

West, G. B., J. H. Brown, and B. J. Enquist. 1997. "A general model for the origin of allometric scaling laws in biology." Science 276 (5309): 122–126.

——. 1999. "The fourth dimension of life: Fractal geometry and allometric scaling of organisms." Science 284:1677–1679.

Westoby, M. 1998. "A leaf-height-seed (LHS) plant ecology strategy scheme." Plant Soil 199:213–227.

Westoby, M., D. S. Falster, A. T. Moles, P. A. Vesk, and I. J. Wright. 2002. "Plant ecological strategies: Some leading dimensions of variation between species." Annual Review of Ecology and Systematics 33:125–159.

Witman, J. D., and K. Roy. 2009. Marine macroecology. Chicago: University of Chicago Press.

# Contributors

Gary M. Barker, Landcare Research, Hamilton, New Zealand

James H. Brown, Department of Biology, University of New Mexico, Albuquerque, New Mexico

Robert A. D. Cameron, Department of Animal and Plant Sciences, University of Sheffield, Sheffield, United Kingdom

Steven L. Chown, Centre for Invasion Biology, Stellenbosch University, Matieland, South Africa

S. K. Morgan Ernest, Department of Biology and the Ecology Center, Utah State University, Logan, Utah

Kevin J. Gaston, Environment and Sustainability Institute, University of Exeter, Penryn, Cornwall, United Kingdom

Kate E. Jones, Institute of Zoology, Zoological Society of London, London, United Kingdom

Astrid Kodric-Brown, Department of Biology, University of New Mexico, Albuquerque, New Mexico

S. Kathleen Lyons, Department of Paleobiology, National Museum of Natural History, Smithsonian Institution, Washington, DC

Pablo A. Marquet, Center for Advanced Studies in Ecology and Biodiversity (CASEB) and Institute of Ecology and Biodiversity (IEB), Departamento de Ecología, Pontificia Universidad Católica de Chile, Santiago, Chile

Brian A. Maurer, Department of Fisheries and Wildlife, Michigan State University, East Lansing, Michigan

Shai Meiri, Department of Zoology, Tel Aviv University, Tel Aviv, Israel

Jeffrey C. Nekola, Department of Biology, University of New Mexico, Albuquerque, New Mexico

Beata M. Pokryszko, Museum of Natural History, Wrocław University, Wrocław, Poland

Kamran Safi, Institute of Zoology, Zoological Society of London, London,

United Kingdom, and Max Planck Institute for Ornithology, Radolfzell, Germany

Felisa A. Smith, Department of Biology, University of New Mexico, Albuquerque, New Mexico

Richard M. Sibly, School of Biological Sciences, University of Reading, Reading, United Kingdom

# Index

*Page numbers in italics refer to tables or figures.*